Analysis and Design of Wood Structures

Comprehensive Design Project

(Can be used for PE exam preparation)

———————

Arzhang Zamani, Ph.D.

STRUCTURE GATE

Analysis and Design of Wood Structures
Copyright © 2017 by STRUCTURE GATE

ISBN (978-1-948135-03-0)

Library of Congress Control Number 2017916020

Printed in the USA

Disclaimer

Dedication

To Our parents, Farzaneh and Mehrdad

With Love and Passion

Table of Contents

List of Tables

Introduction

The purpose of this book is to familiarize the reader with the design and analysis of low-rise wood buildings. Moreover, the reader will gain sufficient experience for successful performance in PE and SE exams held in the USA for licensing of engineers.

This book tries to cover most essential parameters for analysis and design of wooden structures. Therefore, in each chapter design requirements are followed by an example. One of the goals of this book is present the process of applying theories to a real design project. The last section of this book offers a complete structural design of wooden buildings with supporting hand calculations.

I hope you enjoy reading this book.

Live with Passion!

Arzhang Zamani

Chapter 1
Loads and Structure of Wood

1.1 Dead Loads

1- Dead loads are long-term stationary forces that include the self-weight of the structure and the weight of any permanent equipment tributary to the design of a given member.

2- The area assumed to load any given member is called tributary area.

3- How to specify member weights:

- For weight of wood framing (Example: 2x4 @ 12in. o.c. (on center); w=1.2 psf)

- For weight of roof and Floor materials (Example: 1in. thick Plaster; w=8.0 psf)

1.2 Dressed Lumber Size and Weight

The lumber is surfaced to the standard nominal sizes. Planning machine is used to achieve smooth and uniform sizes. It is an important operation so the wood can be used effectively as a construction material; to achieve a uniform, free of imperfections construction process. This operation will also avoid squeaking of floors or rattling of walls while walking or closing doors, respectively.

Note for Sizes: Nominal sizes are reduced for the first and second number

- Members that are 1x reduce the nominal size by ¼". (For example, 1x = ¾"x)

- Members greater than 2x reduce the nominal size by ½". (For example, 2x = 1½"x, 4x = 3½"x, and 18x = 17½"x)

- Members that are greater than or equal to Xx8 but only up to 4x reduce the nominal size by ¾". (For example, 3x8 = 2½" x 7¼", 4x8 = 3½" x 7¼", 4x16 = 3½"x 15¼", and 6x8 = 5½" x 7½")

- Members that are greater than or equal to 5x5 reduce the size by ½".

- Rough *Sawn:* For sizes add 1/8" to those values obtained for dressed lumber.

- *Full Sawn*: No longer used or available in the market unless by special order

Density of Wood

$$\rho = 62.4\left(\frac{G}{1 + G(0.009)(mc)}\right)\left(1 + \frac{mc}{100}\right)$$

ρ: Density of wood (pcf/ft)

G: Specific gravity of wood

mc: Moisture content percent

The weight of wood:

$$w = \rho\left(\frac{A}{144}\right)\left(\frac{12}{S}\right)$$

w: Weight of wood

A: Area in (in^2) of wood member

S: Spacing of joists, rafters, and or purlins in (in.).

Example 1:

Compute weight in psf for a 15% moisture content 2x8 DF-L @ 16" o.c.

Notes: 1- 15% is very typical of wood in construction. 2- DF-L Stands for Douglas Fir – Larch mostly used in home construction. 3- If type is not specified always use Dressed Sizes 4- Density of wood is in pcf per foot of spacing

Arzhang Zamani

Solution:

Standard Dressed size for a 2x8 is 1 ½" x 7 ¼".

$$\rho = 62.4\left(\frac{0.5}{1 + G(0.009)(15)}\right)\left(1 + \frac{15}{100}\right) = 33.6 \text{ pcf/ft}$$

Weight of 2x8 @ 16" o.c.:

$$w = 33.6\left(\frac{1.5(7.25)}{144}\right)\left(\frac{12}{16}\right) = 1.9 \text{ psf}$$

Example 2:

Compute floor and ceiling weight in pcf.

Floor Dead Load:

Carpet Flooring (given)	4.0	psf
5/8" Plywood (B.3)	3 x 5/8	psf
2x8 @ 12 Floor Framing (A)	2.5	psf
R-19 Insulation (loose)	0.5	psf
Misc. (HVAC Equip.)	6.0	psf
Total	14.9	psf

Ceiling Dead Load:

Acoustic Tile	1 x 3/8	psf
2x4 @ 12	1.2	psf
Misc. (Light Fixture)	6.0	psf
Total	8.6	psf

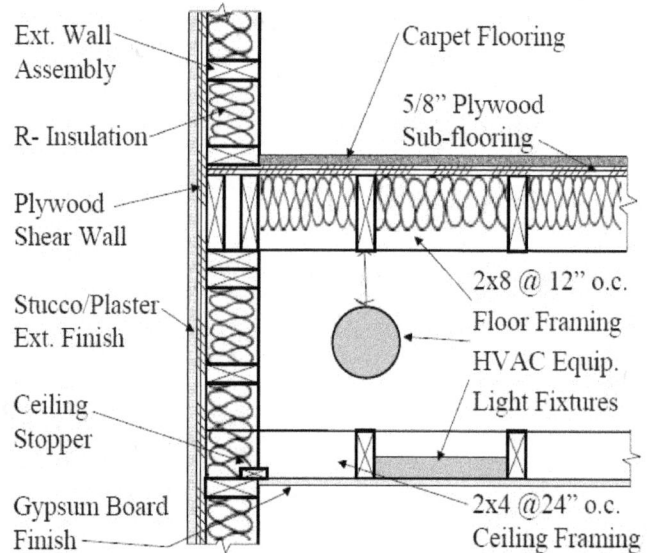

Ext. Wall Assembly
R- Insulation
Plywood Shear Wall
Stucco/Plaster Ext. Finish
Ceiling Stopper
Gypsum Board Finish

Carpet Flooring
5/8" Plywood Sub-flooring
2x8 @ 12" o.c. Floor Framing
HVAC Equip.
Light Fixtures
2x4 @ 24" o.c. Ceiling Framing

4

Table 1 Specific gravity of wood (NDS 2015)

Specific Gravity of Western Softwood Species	
Western Species	Specific gravity (Oven Dry Weight/Oven Dry Volume)
DOUGLAS FIR-LARCH Douglas Fir Western Larch	0.5
DOUGLAS FIR-SOUTH Douglas Fir-South	0.46
HEM-FIR Western Hemlock Noble Fir California Red Fir Grand Fir Pacific Silver Fir White Fir	0.43
SPIRUCE-PINE-FIR (SOUTH) Engelmann Spruce Sitka Spruce	0.36
SPIRUCE-PINE-FIR (SOUTH) MSR 1.2 E to 1.9 E	0.42
MSR 2 E and higher	0.5
ENGLEMANN SPRUCE-LODGEPOLE PINE Engelmann Spruce Lodgepole Pine	0.38
ENGLEMANN SPRUCE-LODGEPOLE PINE MSR 1.5 E and higher	0.46
WESTERN CEDARS Any of the species in the first four species groups above plus any or all of the followings Idaho White Pine Ponderosa Pine Sugar Pine Alpine Fir Mountain Hemlock	0.36

1.3 Live Loads, Snow load, Wind load, and Earthquake load

Please refer the latest version of International Building Code and ASCE 07. These calculations are general for all types structures. Therefore, only application of using them in a design project is presented.

Example 3:

Given the following weights compute the unit length Dead Load at the base of the walls. Compute the unit length Live Load at the base of the walls. (please make all necessary assumptions and do not violate given information)

Asphalt shingles w=2.0 psf

5 ½" Fiberglass loose insulation w= 2.8 psf

3/8" Plywood w=1.1 psf

Reroofing w= 1.5 psf

½" Gypsum Wall Board w=2.5 psf

7/8" Stucco 10.0 psf

Use Douglas Fir-Larch with a 15% moisture content (m.c.)

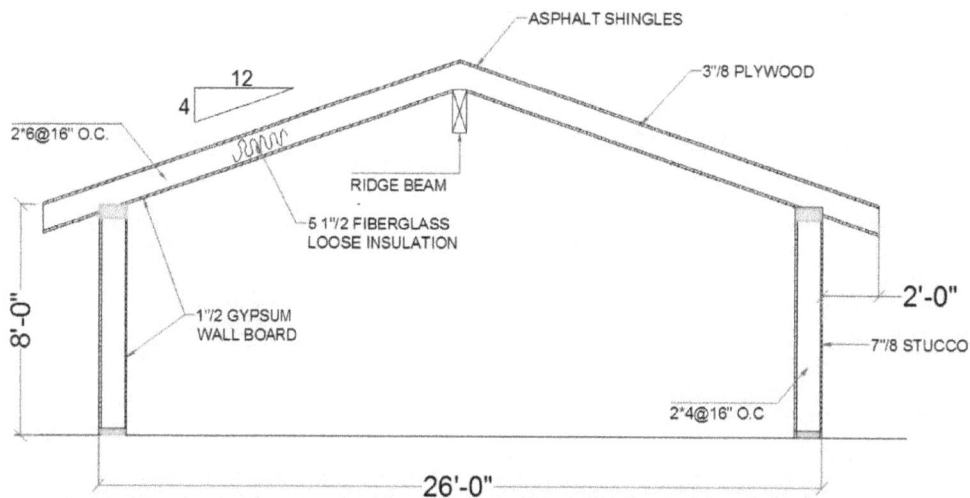

Example 3 Solution:

a) Unit length Dead Load at the base of the walls

This one story elevation is in complete symmetry. Therefore, for calculating the load per wall, only half of frame is evaluated.

Breakdown of Dead Load:

Length of rafter: $\sqrt{225 + 25} = 15.81 \: ft$

Length of rafter to the wall: $\sqrt{169 + 9} = 13.34 \: ft$

Height of wall: $8 \: ft$

$$G = 0.5 \: \& \: mc = 15$$

$$\rho = 62.4 \left(\frac{G}{1 + G(0.009)(mc)} \right) \left(1 + \frac{mc}{100} \right)$$

$$\rho = 33.611 \: pcf$$

Wall:

$$A_{wall} = 5.25 \: in^2 \: \: \& \: \: S = 16 \: in.$$

$$W_{wall} = \rho. A_{wall}. \frac{12}{S} = 1.444 \: plf$$

Dead Load			
Type	Intensity (psf)	Applicable length (ft)	load per square length (plf)
Asphalt Shingles	2	15.8	31.6
5.5" Fiberglass	2.8	15.8	44.2
3/8" Plywood	1.1	15.8	17.4
Reroofing	1.5	15.8	23.7
0.5" Gypsum(vertical)	2.5	8.0	20.0
0.5" Gypsum(roof)	2.5	13.7	34.3
7/8" Stucco	10	8.0	80.0
Wall	0.919	8.0	7.4
Rafter	1.444	15.8	22.8
		Total	281.4

Therefore, the sum of dead load is: 281.4 plf

b) Live Load:

Intensity: 16 psf

Note: Unit live load specified in the code are applied on a horizontal plane
Horizontal length: 15 ft

$$W_L = 15(16) = 240 \; plf$$

Example 4:

According to the below figure

a) Develop the applicable wind forces for a one-story, box-type industrial building, with $q_s = 126.5\ psf$

$C_e \Rightarrow$	Exposure $B \rightarrow \begin{cases} 0.62 & (0' \le h \le 15') \\ 0.67 & (15' \le h \ge 20') \end{cases}$ (UBC Table16 – G)
$C_q \Rightarrow$	$\begin{cases} 0.80 & windward \\ -0.50 & leeward \\ -0.70 & roof \end{cases}$ Assuming » flat roof and closed building
$I_W \Rightarrow$	1.00

b) Assume walls weighing 35.00 lb/ft, and a roof dead load is 12.50 psf. Given the total Base shear is given by $V = 1.25 \left(\dfrac{W}{R}\right)$, develop the applicable seismic forces for a one-story, box-type industrial building.

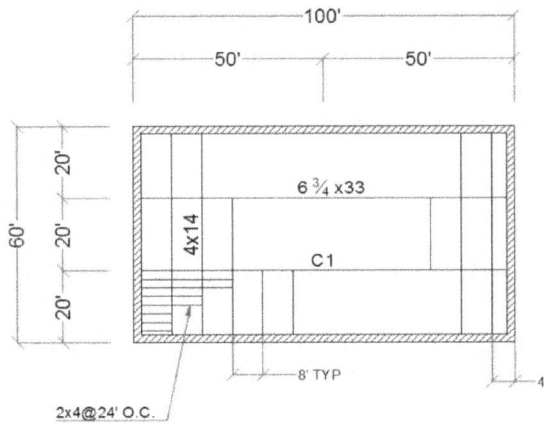

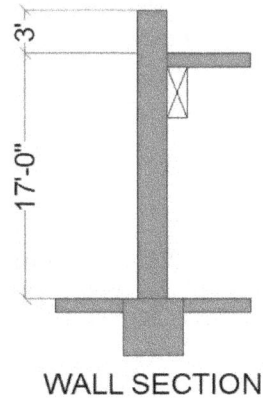

WALL SECTION

Arzhang Zamani

Example 4 Solution:

a) Applicable wind force

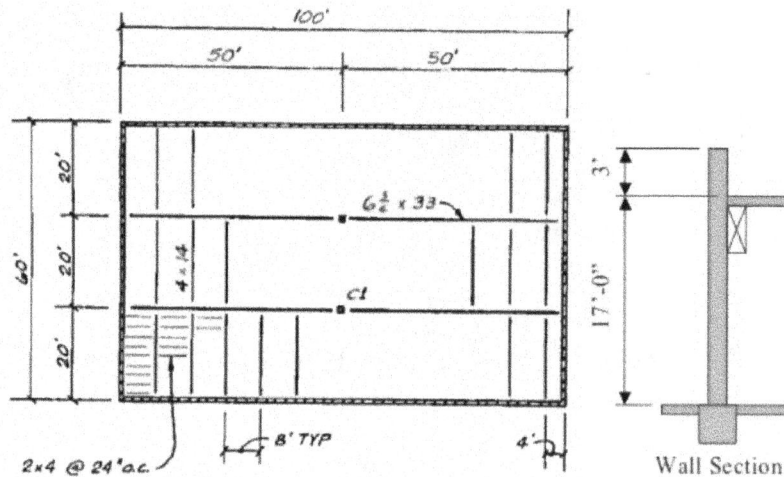

1. $P = C_e . C_q . q_s . I_w$

$$C_e = \begin{cases} 0.62 & 0 < h < 15 \\ 0.67 & 15 < h < 20 \end{cases}$$

$$C_q = \begin{cases} 0.8 & Windward \\ -0.5 & Leeward \\ -0.7 & Roof \end{cases}$$

$q_s = 12.6 \, psf \quad \& \quad I_w = 1$

Note: The transverse and longitudinal direction experience the same wind load

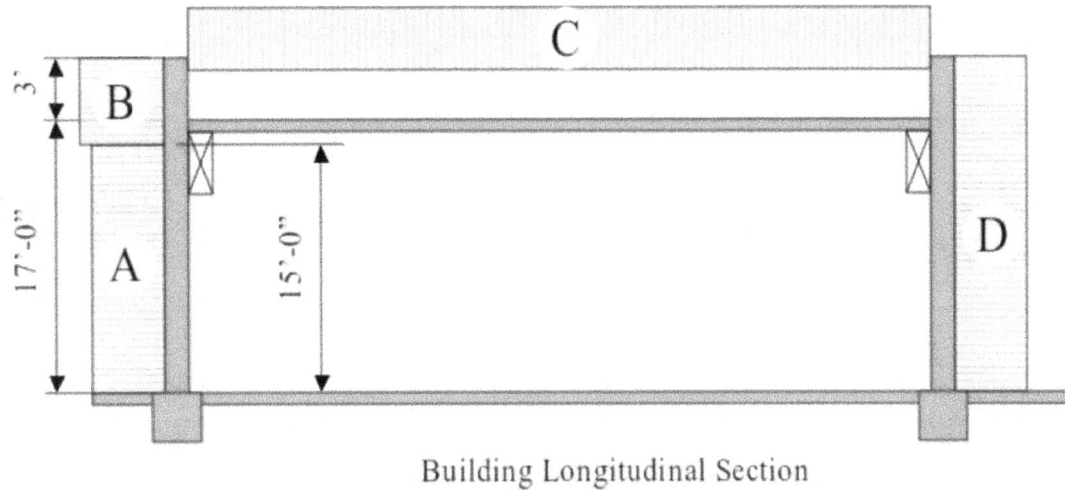

Building Longitudinal Section

Loading. Area for A: $P_A = 0.62(0.8)(12.6)(1) = 6.25 \ psf$

Loading. Area for B: $P_B = 0.67(0.8)(12.6)(1) = 6.754 \ psf$

Loading. Area for C: $P_C = -0.67(0.7)(12.6)(1) = -5.909 \ psf$

Loading. Area for D: $P_D = -0.67(0.5)(12.6)(1) = -4.221 \ psf$

Summing the loads to finding the reaction in roof and base:

$$P_B - P_D = 10.975 \ psf \ \ \& \ \ P_A - P_D = 10.471 \ psf$$

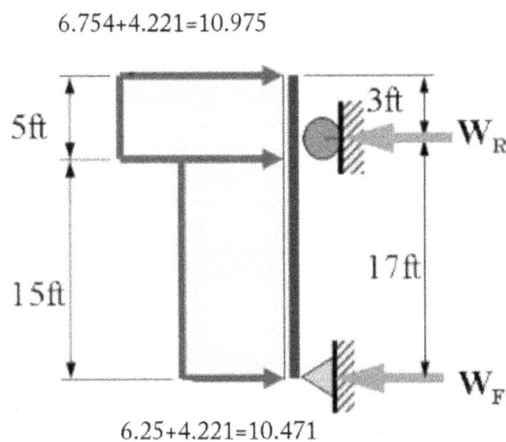

Arzhang Zamani

$$W_F = (P_A - P_D)15\left(\frac{9.5}{17}\right) - (P_B - P_D)5\left(\frac{0.5}{17}\right) = 86.154 \, plf$$

$$W_R = (P_A - P_D)15\left(\frac{7.5}{17}\right) + (P_B - P_D)5\left(\frac{17.5}{17}\right) = 125.778 \, plf$$

Transferring the load to flexible diaphragms:

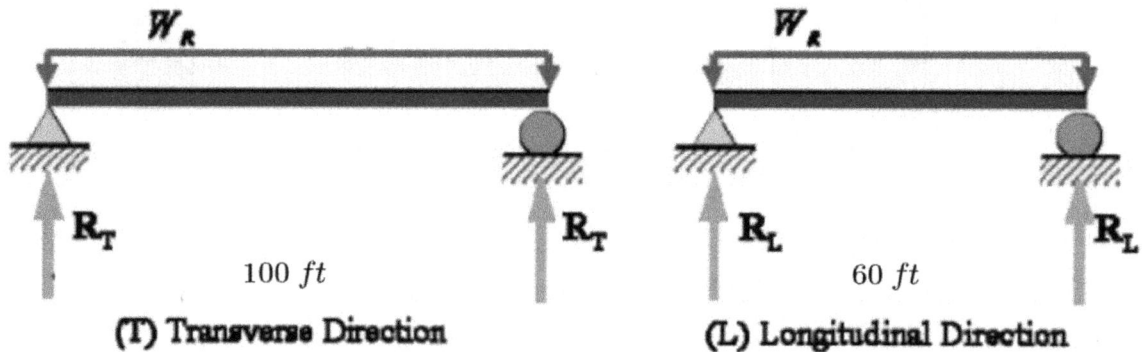

(T) Transverse Direction   **(L) Longitudinal Direction**

Transverse direction:

$$R_T = W_R\left(\frac{100}{2}\right) = 6{,}289 \, lbf$$

Longitudinal direction:

$$R_T = W_R\left(\frac{60}{2}\right) = 3{,}773 \, lbf$$

b) Earthquake load:

$$V = 1.25\left(\frac{W}{R}\right)$$

$$W_{roof} = 12.5 \, psf \quad W_{wall} = 35 \, psf \quad R = 4.5$$

Transverse direction:

Roof: $W_R = W_{roof}(60) = 750 \, plf$

Walls: $W_W = 2(W_{wall})\left(\dfrac{\frac{20^2}{2}}{17}\right) = 823.53\ plf$

$$W_1 = W_R + W_W = 1{,}574\ plf$$

$$V_T = 1.25\left(\dfrac{W_1}{R}\right) = 437.1\ plf$$

Longitudinal direction:

Roof: $W_R = W_{roof}(100) = 1{,}250\ plf$

Walls: $W_W = 2(W_{wall})\left(\dfrac{\frac{20^2}{2}}{17}\right) = 823.53\ plf$

$$W_2 = W_R + W_W = 2{,}074\ plf$$

$$V_L = 1.25\left(\dfrac{W_2}{R}\right) = 575.98\ plf$$

Transverse direction:

$$R_T = V_T(\dfrac{100}{2}) = 21{,}855\ plf$$

Longitudinal direction:

$$R_L = V_L\left(\dfrac{60}{2}\right) = 17{,}279\ plf$$

1.4 Material Properties

VISUALLY GRADED:

Sorted into grades by visual inspection according to growth characteristics. Most lumber is visually graded. The strength of visually graded lumber within a graded class is variable; characterized by large coefficients of variation of approximately 25%.

MACHINE STRESS RATED, MSR:

Nondestructive, automotive measurement of member's stiffness (Modulus of Elasticity) which is correlated to strength properties. Accompanied by an additional visual inspection. MSR lumber has reduced coefficients of variation, in the neighborhood of 11%.

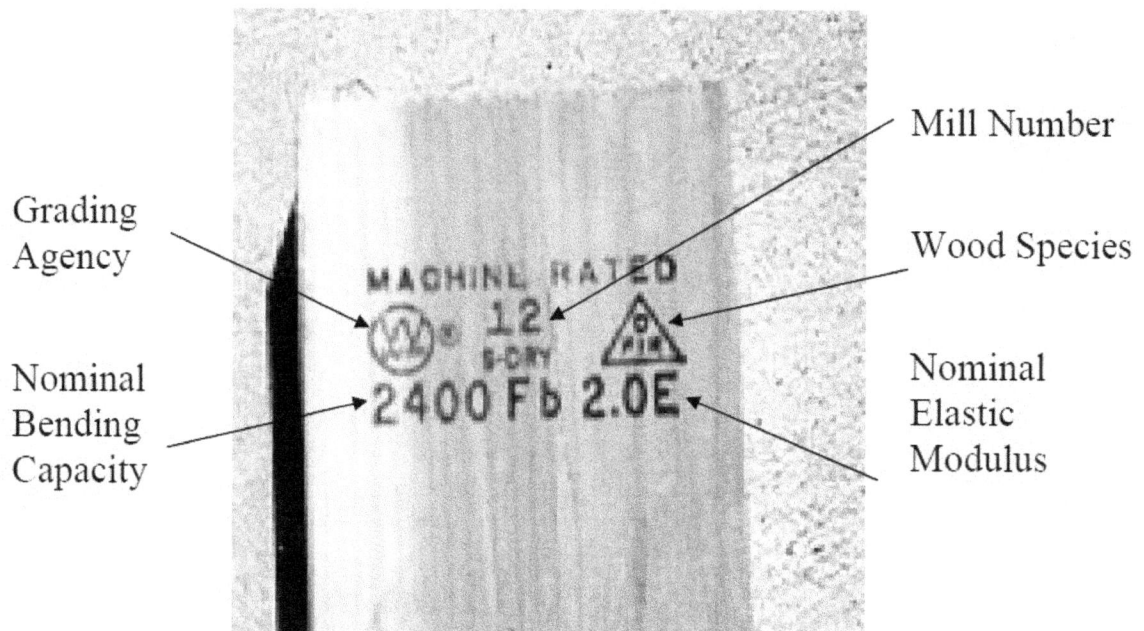

Structural properties, quantified regarding stress grades, are very dependent upon inherent growth characteristics. The characteristics of major importance to engineers are discussed below:

Moisture content (MC) facts:

- In a living tree, MC ≈ 100 to 200%.
- Structural lumber in service, MC ≈ 7 to 14%.
- To calculate MC:

  MC = ([moist weight - oven dry weight] / oven weight) X 100

- The fiber saturation point, FSP, is the moisture content corresponding to a loss of all free water, which is the water contained in cell cavities. Water, however, is still within the cell walls.
- Below the FSP of $\approx$ 30%, decreases in MC are associated with increases in strength peaking at MC $\approx$ 10%.
- S-dry: surfaced dry lumber.
- S-grn: surfaced, unseasoned lumber.
- Assumed in-service MC for S-dry is $\approx$ 15% and for S-grn is $\approx$19%.
- Because of potential in-service moisture content variations, you should carefully consider the effects of both swelling and shrinkage of structural members. Shrinkage is of concern with connections that can restrain the movement of the drying wood.

DENSITY:

In general, denser wood = stronger wood. The more summer wood (latewood) vs. spring wood (earlywood), the denser (stronger) the wood is.

KNOTS:

Act as stress concentrations causing a decrease in strength. Knots displace clear wood. The wood grain must pass around the knot. Lumber grading rules account for knot size, type, and distribution.

In this book, DF-L dimension lumber is commonly used. Table 2 shows the design values for DF-L lumber which is excerpted from Table 4A of the NDS (NDS 2015).

Table 2 Structural design values for DF-L lumber (NDS 2015)

Base Design values for Visually Graded DF-L Dimension Lumber						
Douglas Fir-larch	Design values in pounds per square inch (psi)					
	Bending F_b	Tension parallel to grain F_t	Shear parallel to grain F_v	Compression perpendicular to grain $F_{c\perp}$	Compression parallel to grain F_c	Modulus of Elasticity E
Select Structural	1500	1000	95	625	1700	1,900,000
No. 1 and Btr	1200	800	95	625	1550	1,800,000
No. 1	1000	675	95	625	1500	1,700,000
No. 2	900	575	95	265	1350	1,600,000
No. 3	525	325	95	625	775	1,400,000
Stud	700	450	95	625	850	1,400,000

1.5 Adjustment Factors

Introduction to the various adjustment factors required to account for the behavior of structural wood members. Emphasis is given to those factors that apply to wood beams.

- The structural behavior of wood is affected by the many in-service conditions like humidity, temperature, and relative member size.
- These effects are accounted for in ASD by applying adjustment factors to the base design values tabulated in the NDS Supplement.
- A prime notation (i.e. ') is used here to indicate that the stated allowable stress has been adjusted. In the example, consider bending:
- For beam design, we should examine the following states:

1- Bending. F_b'

2- Shear. F_v'

3- Deflection. E'

4- Compression perpendicular to the grain at bearing points. $F_{c\perp}'$

   - The related allowable stress adjusted equations for simple SOLID-SAWN BEAMS in common applications are: (NDS Table 4.3.1)

1. $F_b' = F_b \, (C_D C_M C_t C_L C_F C_{fu} C_r)$

2. $F_v' = F_v \, (C_D C_M C_t C_H)$

3. $E' = E(C_M C_t)$

4. Compression perpendicular to the grain at bearing points. $F_{c\perp}'$

- The bending equation for GLULAMS is:

$$F_{bx}' = Min(F_b \, (C_D C_M C_t C_V), \, F_b \, (C_D C_M C_t C_L))$$

- Unless specifically identified, it will be assumed that bending occurs about the strong axis as shown in the figure below.

1- Strong axis is the principal axis with the largest moment of inertia.

2- The strong axis is labeled the x-axis here with the corresponding moment of inertia, Ix.

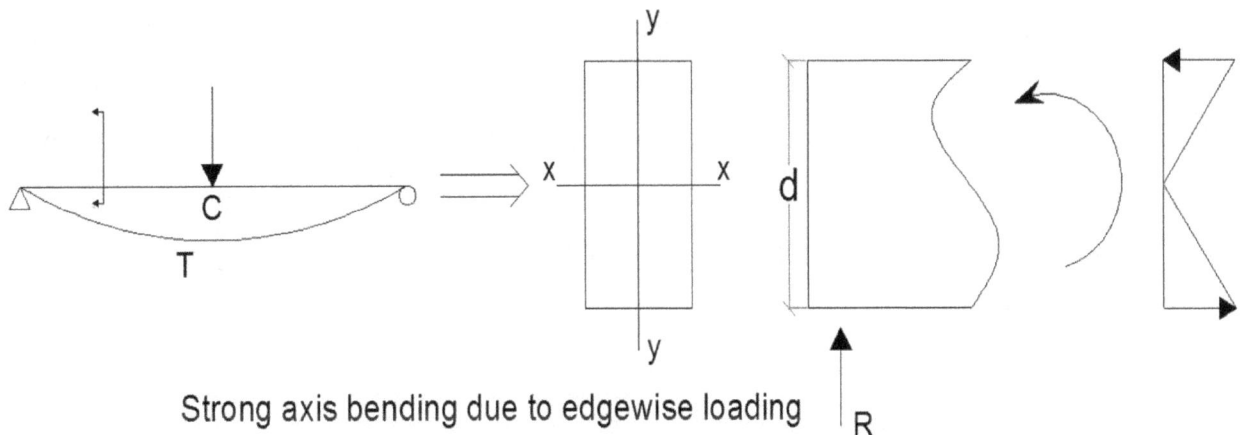

Strong axis bending due to edgewise loading

ADJUSTMENT FACTORS:

Load Duration Factor, C_D (NDS Table 2.3.2)

1. Table 4A base design values apply only to normal load duration.

2. Wood may support higher stress if the load is supported for a short period.

3. Overloads are most likely a result of temporary (short duration) loads.

4. Load duration factor C_D converts normal duration design values for other load duration values.

5. Normal duration load is typically taken as ten years.

6. Floor live loads are within the 10-year duration value.

Table 3 Frequently used load duration factors (NDS Table 2.3.3)

Load Duration	C_D	Typical Design Loads
Permanent	0.9	Dead
10 Years	1	Occupancy Live
2 Months	1.15	Snow
7 Days	1.25	Constriction
10 Minutes	1.6	Wind/Earthquake
Instant	2	Impact

In beam design, however, engineers design for combinations of loads where each load of the combination has a different duration. According to the NDS, the shortest duration load in a combination of loads controls the selection of C_D. And, many times engineers need to check more than one load case. Here is the example for determining which load case, with its associated C_D, will be the critical design load combination.

Example 5:

Consider a fully laterally supported bending member subject to tributary dead load of D=80 plf, roof snow load of S=320 plf, wind uplift of W=±128 plf, and earthquake of E=±180plf. Determine the critical load combination for this bending member. The member only supports the roof above. The following un-factored load cases for C_D are applicable:

D	+80
D + L	+80
D + L + (L_R or S or R)	80 + 320 = +400
	+W: 80 + 128 + 320 = +528
D + (W or 0.7E) + L + (L_R or S or R)	-W: 80 - 128 + 320 = +272
	+E: 80 + 0.7x180 + 320 = +526
	-E: 80 - 0.7x180 + 320 = +274
0.6D + W	+W: 0.6 x 80 + 128 = +176
	-W: 0.6 x 80 - 128 = -80
0.6D + 0.7E	+E: 0.6 x 80 + 0.7x180 = +174
	-E: 0.60 x 80 - 0.7x180 = -78

Using the straight values from above the design loads that will govern are +528 plf and -80 plf.

The following factored load cases for C_D are applicable:

D	$C_D = 0.90$	+80 / 0.90 = +89
D + L	$C_D = 1.00$	+80 / 1.00 = +80
D + L + (L_R or S or R)	$C_D = 1.15$	+400 / 1.15 = +348
		+W: +528 / 1.60 = +330
		-W: +272
D + (W or 0.7E) + L + (L_R or S or R)	$C_D = 1.60$	+E: +526
		-E: +274
0.6D + W	$C_D = 1.60$	+W: +176
		-W: -80
0.6D + 0.7E	$C_D = 1.60$	+E: +174
		-E: -78

Size Factor, C_F (NDS 4.3.6)

The relative size (width, depth, and length) of a wood member has an effect on the member's strength, which is accounted for by C_F for solid sawn members and C_V for glued laminated timbers. the bigger a member gets and/or the longer its maximum stressed region is, the greater the probability of a weak link occurrence with reduced strength.

Table 4 Size factor for solid sawn members (NDS 2015)

Grades	Width	F_b		F_t	F_c
		Thickness			
		2" & 3"	4"		
Select Structural No. 1 & Btr. No. 1, No.2, No. 3	2", 3" & 4"	1.5	1.5	1.5	1.15
	5"	1.4	14.0	1.4	1.10
	6"	1.3	1.3	1.3	1.10
	8"	1.2	1.3	1.2	1.05
	10"	1.1	1.2	1.1	1.00
	12"	1.0	1.1	1.0	1.00
	14" & wider	0.9	1.0	0.9	0.90
Stud	2",3" & 4"	1.1	1.1	1.1	1.05
	5" & 6"	1.0	1.0	1.0	1.0
Constriction & Standard	2" , 3" & 4"	1.0	1.0	1.0	1.0
Utility	4"	1.0	1.0	1.0	1.0
	2" & 3"	0.4	-	0.4	0.6

TIMBER:

This value is not used for glulam. For glulams must use the volume factor instead.

$$C_F = \left(\frac{12}{d}\right)^{\frac{1}{9}} \leq 1.0$$

- The size factor for Southern Pine dimension lumber is incorporated directly, for the most part, into the Southern Pine design value table, Table 4B.
- This is in fact why there is a different table for Southern Pine distinct from the other commercial species grouping.

- There are some exceptions, however, and it is necessary to apply a C_F factor for the following cases:

1- For dimension lumber 4" thick and $\geq$ 8" in width for all grades except Dense Structural 86, Dense Structural 72, and Dense Structural 65: multiply F_b by a C_F factor of 1.1.

2- For dimension lumber $\geq$ 12".

3- For 12" wide members for all grades except Dense Structural 86, Dense Structural 72, and Dense Structural 65: multiply F_b, F_t, and F_c by a C_F factor of 0.9.

4- For Dense Structural 86, Dense Structural 72, and Dense Structural 65 members with depth greater than 12", multiply F_b by a C_F factor obtained from the following:

$$C_F = \left(\frac{12}{d}\right)^{\frac{1}{9}} \leq 1.0$$

Example 6:

Find the size factor for the following members:

Visually graded members:

2x4 →1.5   4x2 →1.5   2x6 →1.3

4x4 →1.5   4x8 →1.3   4x10 →1.2

Repetition Member Factor, C_r (NDS 4.3.9)

For dimension lumber (2" to 4" thick) systems used in a closely-spaced and repetitive manner, the bending design value may be adjusted upward by C_r =1.15.

This adjustment may be taken if all the following criteria apply:

1- Dimension lumber members are used as joists, truss chords, rafters, studs, planks, or decking.

2- The members are spaced $\leq$ 24" on center.

3- There are 3 or more parallel members in the system.

4- The members are joined by some load distribution system like sheathing, subflooring, flooring, or tongue and groove joints.

Example 7:

Find the repetition factor for the following members:

2x4@12"o.c →**1.15**   2x4@16"o.c →**1.51**

2x4@24"o.c →**1.15**   4x10@48"o.c →**1.00**

Flat Use Factor, C_{fu} (NDS 4.3.7)

It is assumed in the NDS basic design value tables, that bending members are stressed in flexure about their strong axis, the x-axis. This type of loading is known as edge-wise or load applied to the narrow face. Sometimes, however, wood members may be loaded in bending about the weak axis, known as the y-axis. When this type of loading occurs, you may be able to increase the appropriate tabulated bending stress value by C_{fu} depending upon member type as outlined below:

- For solid sawn dimension lumber (2" to 4" thick), any species, visual or machine graded, F_b may be adjusted by the following C_{fu} factors.

Table 5 Flat use factors, NDS table 4A (NDS 2015)

Width	Thickness	
	2" & 3"	4"
2" & 3"	1	-
4"	1.1	1
5"	1.1	1.05
6"	1.15	1.05
8"	1.15	1.05
10" & wider	1.2	1.1

(Header: FLAT USE FRACTORS, C_{fu})

Wet Service Factor, CM (NDS 4.3.3)

Lumber is graded at 19% EMC - Equilibrium Moisture Content (EMC)

- With respect to the in-service moisture content of the wood members, the tabulated NDS design values are specified according to the following assumptions:

1- Lumber is used in dry conditions like that found in most covered structures except for swimming pool enclosures or moist industrial environments.

2- The moisture content of solid sawn wood member in service is $\leq 19\%$, and moisture content for glued laminated timbers is $\leq 16\%$, regardless of the moisture content at the time of manufacture.

3- If the in-service wood moisture content exceeds 19% for solid-sawn members and 16% for glulams for an extended period, contradicting the above assumptions, then the allowable stresses need to be adjusted by CM, as given in the tables below.

Solid Sawn Dimension Lumber including Visually Graded, MSR, and Southern Pine:

Table 6 Wet service factor (NDS 2015)

When dimension lumber is used where moisture content will exceed 19% for an extended time period, design values shall be multiplied by appropriate wet service factor form the following table:

WET SERVICES FACTORS,C_M					
F_b	F_t	F_v	$F_{c\perp}$	F_c	E
0.85*	1.0	0.97	0.67	0.8**	0.9

* when $(F_b)(C_F) \leq 1150$ psi, $C_M = 1.0$

** when $(F_c)(C_F) \leq 750$ psi, $C_M = 1.0$

Temperature Factor, C_t (NDS 4.3.4)

Wood strength decreases with sustained exposure to higher than normal temperatures. This temperature effect is immediate, and magnitude varies as a function of wood moisture content. Wet wood experiences greater temperature induced reductions in strength. Up to 150° F, the temperature effect is reversible; member recovers all its strength upon return to normal temperatures. In general, these elevated temperatures occur simultaneously with low wood moisture contents resulting in offsetting strength effects and elevated temperature is short term. Therefore, its traditional practice in these areas to use Ct of 1.0.

Account for solid-sawn or glulam situations where the in-service ambient temperature exceeds 100°F for extended periods of time by applying Ct as given in the following table.

Table 7 Temperature Factor values, NDS table 2.3.3 (NDS 2015)

Design Values	In-service Condition	T ≤ 100° F	100°F ≤ T ≤ 125° F	125°F ≤ T ≤ 150° F
F_t, E	Wet or Dry	1.0	0.9	0.9
F_b, F_v, $F_{c\perp}$, F_c	Dry	1.0	0.8	0.7
F_b, F_v, $F_{c\perp}$, F_c	Wet or Dry	1.0	0.7	0.5

1.6 Structural Glued Laminated Timber

The term glue-laminated timber refers to an engineered, the stress-rated product comprising parallel assemblies of wood laminations, finger-jointed at the ends, and face laminated with exterior-grade adhesives.

There are many benefits associated to Glue laminated timber which is economical, creative, Conserves resources, Kiln-dried, waterproof, insulating, and use of renewable resources.

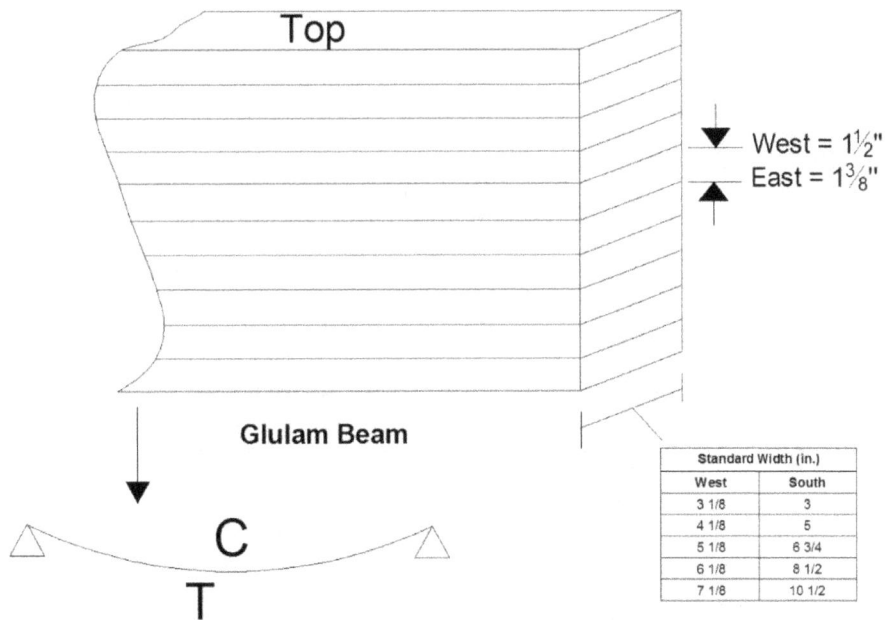

Two main glulam grades:

- Bending combinations
    1- Stressed primarily in bending.

2- Dispersal of various grades of laminates with higher quality laminations located at the top and bottom of the beam.

3- The assumption is that top laminations are in compression and bottom laminates are in tension.

4- Table 5A in NDS: Base Design Values for Structural Glued Laminated Softwood Timber (stressed in bending).

- Axial combinations

1- Assumed to be loaded axially.

2- Uniform distribution of laminates.

3- Specified according to combination symbol and species. e.g. 20-HF

4- Table 5B in NDS: Design Values for Structural Glued Laminated Softwood Timber (stressed axially).

Volume Factor, C_V

Size effect (C_F) is accounted for in glued laminated timbers as a volume effect, C_V. (Table 5A Adjustment factors)

- It incorporates not only cross-section size but also the length of span in a maximum bending stress condition.

- It is applied only to F_{bxx}, (bending about the strong axis)

- Tabulated values are for w = 5-1/8", d = 12" and L = 21'. For other values must use the volume factor C_V.

- Not to be applied simultaneously with the stability factor C_L.

- The volume factor is obtained from the following equation.

$$C_V = k \left(\frac{12}{d}\right)^{\frac{1}{x}} \left(\frac{5.125}{b}\right)^{\frac{1}{y}} \left(\frac{21}{L}\right)^{\frac{1}{z}} \leq 1.0$$

Where b = width ≤ 10-3/4"

d = beam depth

L = span length, in feet, between zero moment points.

x = y = z = 10 for western species.

x = y = z = 20 for Southern Pine.

K is given by the following diagrams:

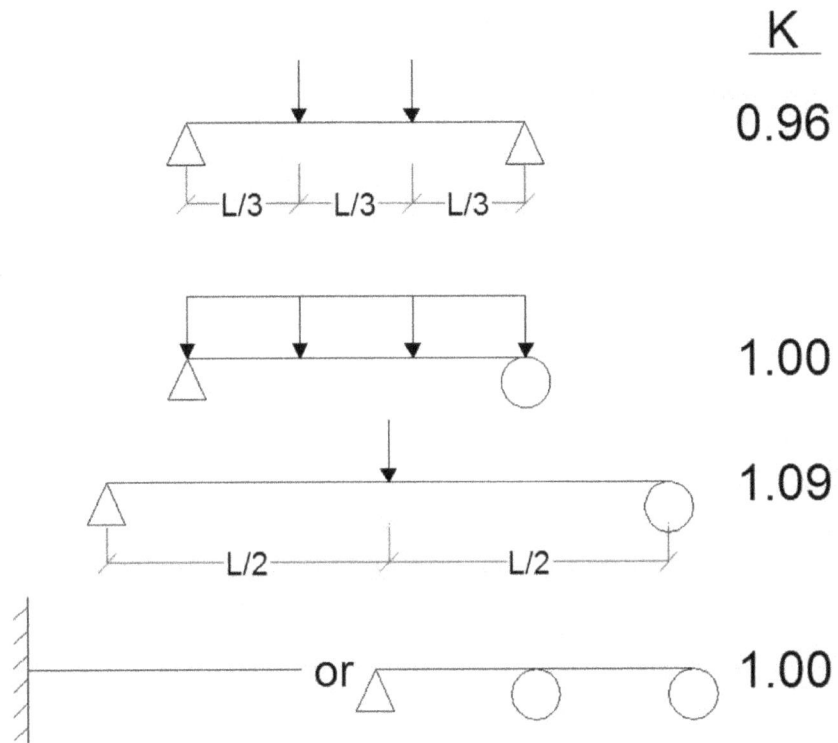

K

0.96

1.00

1.09

or

1.00

Chapter 2
Design of Wood Beams

The design of rectangular Sawn and glulam beams is covered in this lecture. The design of wood beams cover the following:

1	Bending	$F_b' = F_b\,(C_D\ C_M\ C_t\ C_L\ C_F\ C_{fu}\ C_r)$
2	Shear	$F_v' = F_v\,(C_D\ C_M\ C_t\ C_H)$
3	Deflection	$E' = E\,(C_M\ C_t)$
4	Bearing	$F_{c\perp}' = F_{c\perp}\,(C_M\ C_t\ C_b)$
5	Connections	See figures next pages

- F_b: Bending (beam bending)
- F_t: Tension (hanging something from timber, truss member under tension)
- F_v: Shear parallel to grain (Beam design)
- $F_{c\perp}$: Compression perpendicular to grain (Beam bearing)
- Fc: Compression parallel studs in a wall to grain (header support, truss chord in compression)
- E is Modulus of Elasticity (displacement or column buckling calculations)

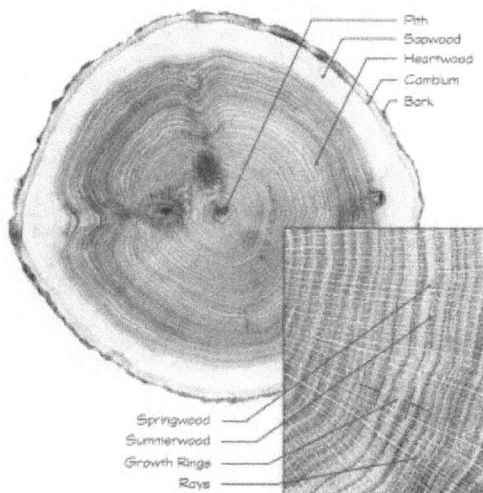

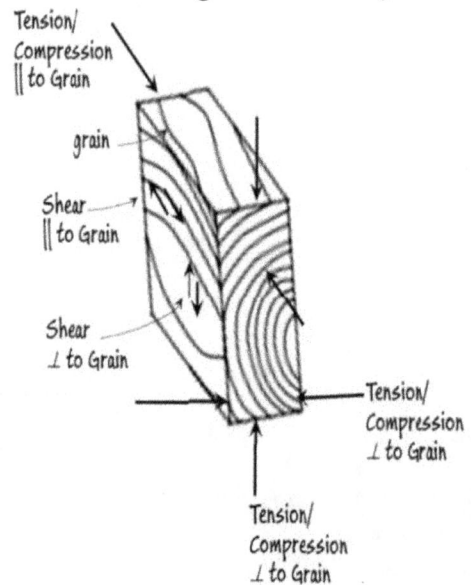

CONNECTIONS IN WOOD BEAMS:

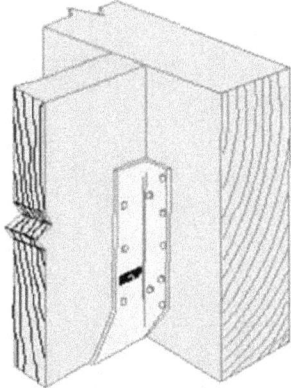

Joist Connection to Beam

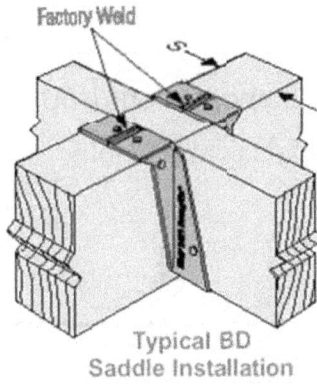

Beam Connection to Beams

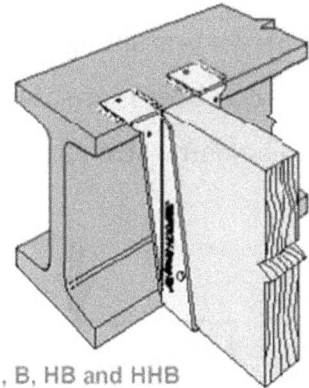

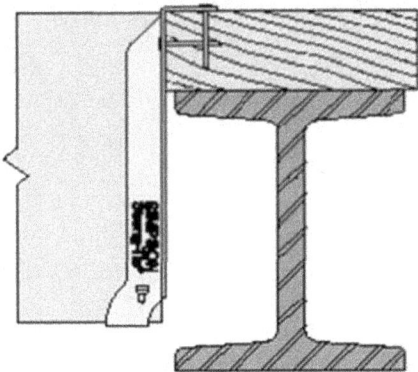

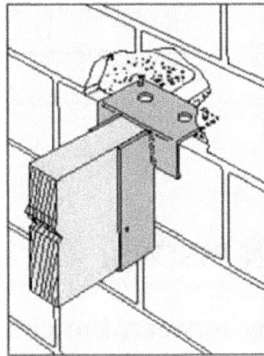

Beam Connection to Beams

Beam Connection to Beam

Beam Connection to Wall

Beam Connection to Post

2.1 Bending in Beams

Assumptions in the bending of beams regarding elastic theory:

1- Plane sections before loading remain plane after loading.
2- Stress is assumed to vary linearly with strain.
3- Wood beams are assumed a homogeneous isotropic material.
4- The Same strength of materials is applicable.

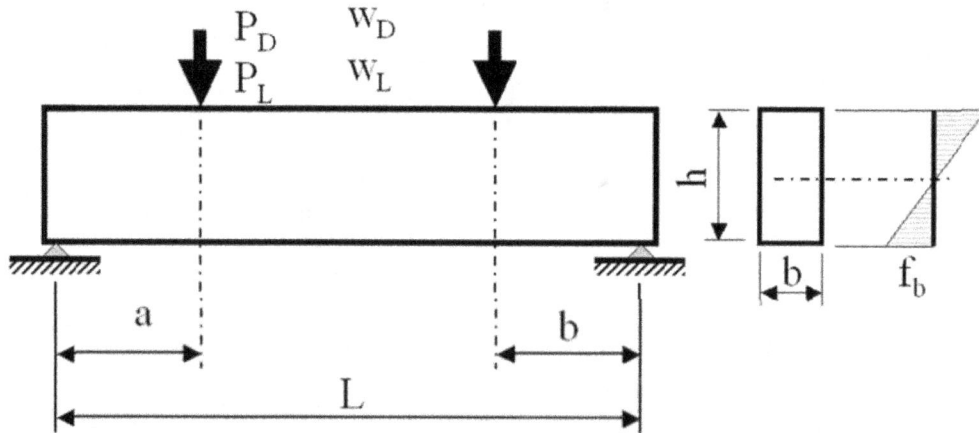

BENDING OF BEAMS DESIGN ISSUES:

1- Span length is the distance between the centerline of required bearing lengths at each end.

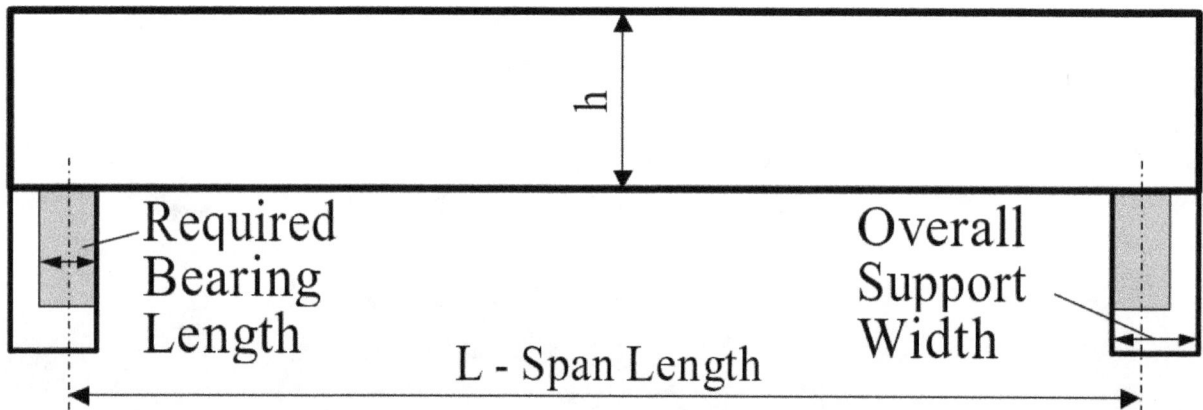

2- Allowable Bending Stresses:

In beam design, one important criterion is to examine the beam's critical bending stress conditions. This approach includes ensuring that the actual critical stress, (e.g. f_b), developed as a function of loading, support conditions, span, and cross-section is ≤ the factored allowable stress, (e.g. F'_b).

The following equation has been specialized for examining the maximum stress situation (at the extreme fibers) for rectangular shaped members.

$$f_b \leq F'_b$$

$$f_b = \frac{Mc}{I} = \frac{M}{S} = \frac{6M}{b\,d^2}$$

$$F'_b = F_b \prod C'_s$$

Arzhang Zamani

Table 4A – Base Design Values for Visually-Graded Dimension Lumber

Note: 1. This table does not include Glu-lams or other engineered products.

2. Must use this table with adjustment factors

Table 8 Structural design values for DF-L lumber (NDS 2015)

Douglas Fir-larch	Bending F_b	Tension parallel to grain F_t	Shear parallel to grain F_v	Compression perpendicular to grain $F_{c\perp}$	Compression parallel to grain F_c	Modulus of Elasticity E
Base Design values for Visually Graded DF-L Dimension Lumber						
Design values in pounds per square inch (psi)						
Select Structural	1500	1000	95	625	1700	1,900,000
No. 1 and Btr	1200	800	95	625	1550	1,800,000
No. 1	1000	675	95	625	1500	1,700,000
No. 2	900	575	95	265	1350	1,600,000
No. 3	525	325	95	625	775	1,400,000
Stud	700	450	95	625	850	1,400,000

Example 8:

Design a solid Sawn wood beam for the given conditions:

1- Use No. 1 DF-L.

2- L_u = 0 ft. (beam is fully braced for out of plane buckling)

3- MC < 19%.

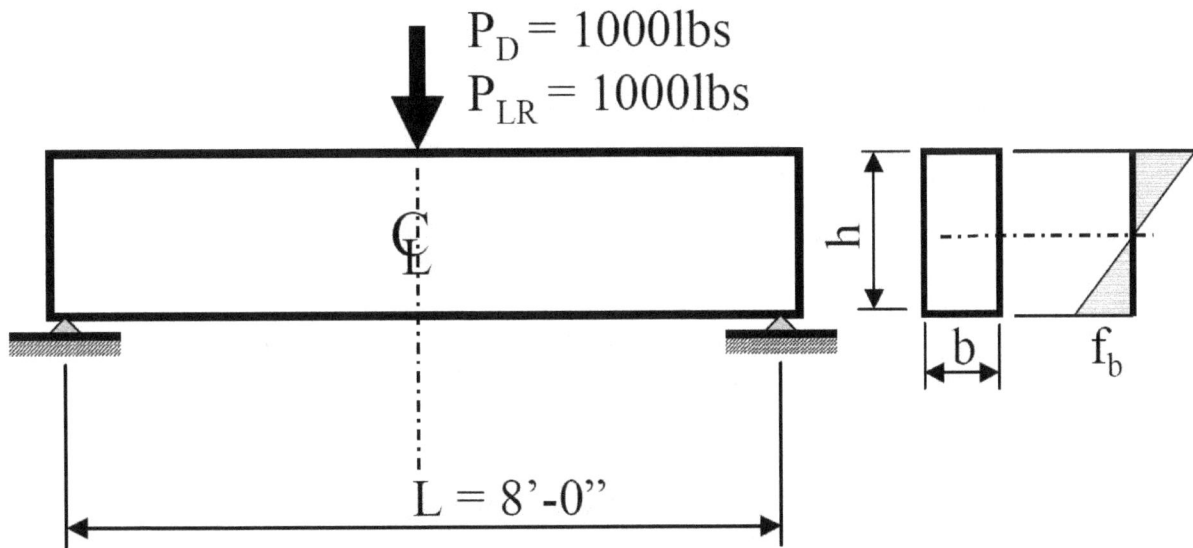

Solution:

1. ASD Load Combinations:

Use duration factor to estimate governing loading case):

D = 1000lbs (neglect beam weight for preliminary calculations).

L = 1000lbs

C_D = 1.25 for D + L_R

C_D = 0.90 for D only

Check governing load case based on C_D:

    a) D = 1000lbs/0.9 =1,111 lbs
    b) D + L_R = (1000+1000) / 1.25 = 1,600 lbs ∴Governs Design

2. Estimate Section Dimension:

Note: assume all other adjustment factors are 1.00 to the exception of C_D=1.25 and C_F = 1.30

    1- M_{max} = PL/4 = $(\frac{2000(8)}{4})(12)$ = 48,000 lbs-in.

    2- F_b =1,000 psi (Table 4A) DF-L No. 1

    3- F_b' = F_b $(C_D C_M C_t C_L C_F C_{fu} C_r)$ = 1000 x 1.25 x 1.30 x 1.00 = 1,625 psi

    4- S_{xx} = $\frac{48,000}{1,625}$ = 29.53 in^3

    5- Try a 4 x 8; $S_{xx(4x8)}$= 30.66 in^3

3. Check Design

$C_F = 1.30, F_b' = F_b \left(C_D C_M C_t C_L C_F C_{fu} C_r \right) = 1,625 \; psi$

$$M_{max} = \frac{DL^2}{8} + \frac{PL}{4}$$

$$\rho = 62.4 \left(\frac{0.5}{1 + G(0.009)(19)} \right) \left(1 + \frac{19}{100} \right) = 34.2 \; \text{pcf/ft}$$

$$D = 6.168 \; plf$$

$$M_{max} = \left(\frac{6.168(8^2)}{8} \right)(12) + 48,000 = 48,592 \; lb - in.$$

$$f_b = \frac{M_{max}}{S_{xx}} = 1,585 \; psi < 1,625 \; psi$$

USE 4 x 8 DF-L No. 1

Example 9:

Design a Glu-Lam beam for the given conditions:

1- Use 24F-V2 SP/SP

2- $L_u = 0$ ft. (beam is fully braced for out of plane buckling)

3- MC < 16%.

4- $W_D = 200 \; plf \; ; W_s = 300 \; plf$

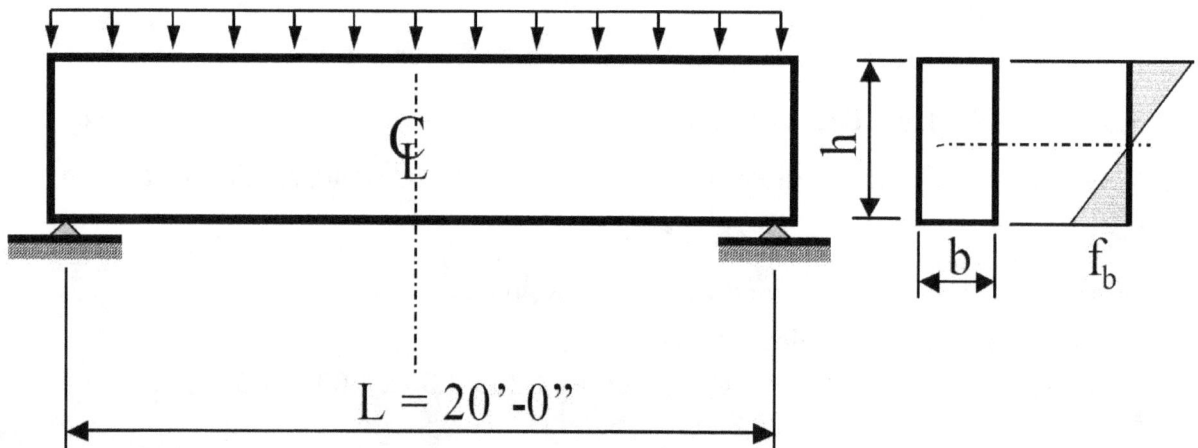

$$L = 20'-0''$$

Solution:

Use duration factor to estimate governing loading case):

D = 200 plf (neglect beam weight for preliminary calculations).

L = 300 plf

C_D = 1.15 for D + S

Check governing load case based on C_D:

a) D = 200/0.9 =222 plf
b) D + S = (200+300) / 1.15 = 434 plf ∴Governs Design

4. Estimate Section Dimension:

Note: assume all other adjustment factors are 1.00 to the exception of C_D=1.15 and C_V = 1.00

1- $M_{max} = \frac{(200+300)(20^2)}{8}(12) = 300,000\ lb - in.$

2- F_b = 2,400 psi

3- $F_b' = 2,400(1.15) = 2,760\ psi$

4- $S_{xx} = \frac{300,000}{2,760} = 109\ in^3$

5- Try a 5 1/8 x 12; since CV = 1.00

Use 5 1/8 x 12 (24F-V1 SP/SP)

2.2 Design of Wood Beams: Lateral Stability

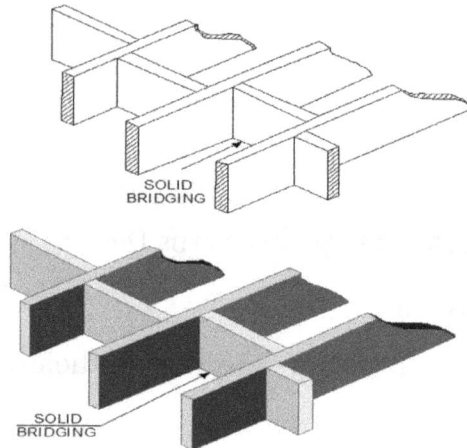

SOLID
BRIDGING

SOLID
BRIDGING

Use duration factor to estimate governing loading case.

Beam stability is a critical design consideration for beams of all materials, including wood beams. Beam stability is a function of the compression stresses that occur due to the bending. The compression fibers of the beam, located along the top half of the simply supported beam with normal downward loading, are susceptible to buckling. Buckling is a sudden failure mode that occurs without warning. Without lateral support or proper design consideration, the compression side of a beam may buckle permanently at a load much smaller than was intended to a normal bending mode.

This load carrying capacity is a function of compression-induced buckling of the beam fibers and is accounted for by either using rule of thumb or by applying a beam stability factor called C_L.

Rule of thumb (Approximate Method):

These rules, which have been used since 1944, apply to only rectangular solid sawn **not** glulam beams. If you provide lateral support according to these rules, then CL essentially becomes equal to 1.0

b = width (inches)
d = depth (inches)

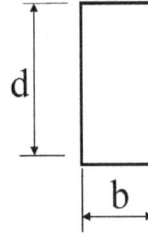

Table 9 Approximate method for providing lateral support

d:b (nominal dimension)	Lateral Support
if d ≤ b	None
2:1	None
3:1 or 4:1	At ends of members by blocking, bridging, hangers.
5:1	The compression edge is held in line the entire length.
6:1	Full depth blocking at 8' o.c.
7:1	Both edges (compression and tension) held in line the entire length of the beam.

The rules should be applied cumulatively. For instance, if you have a 5:1 case, then the lateral stability requirements are:

- The compression edge is held inline the entire beam length,
- The member ends are held in position with full depth blocking, hangers, or bridging.

Beam stability factor C_L

Note: when $d > b$, there is a requirement to provide lateral support at bearing points to prevent rotation or lateral displacement, regardless of the calculated C_L value.

The beam stability factor parallels the treatment given to column buckling. It is a function of the beam slenderness ratio; a measure of the beam's tendency to buckle.

$$R_B = \sqrt{\frac{l_e d}{b^2}} \leq 50$$

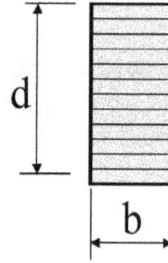

R_B shall not exceed 50.

b = Width (inches)

d = depth (inches)

l_e = Effective span length = function of (end condition, loading, and distance between lateral support points)

Effective length:

- The effective span length accounts for the length of the beam that is unbraced while sustaining large compression stresses.
- It is desirable to have smaller l_e which equals to larger C_L.
- Effective length equations which are given below are a function of term l_u, the unsupported length.
- l_u is the distance between compression edges or bearing bracing points.
- The effective length can be conservatively calculated for cantilever or simply supported beams with any loading according to the following:

$$l_e = \begin{cases} 2.06\, l_u & l_u/d < 7 \\ 1.63\, l_u + 3d & 7 \leq l_u/d \leq 14.3 \\ 1.84\, l_u & l_u/d > 14.3 \end{cases}$$

A more accurate determination of l_e for single span beams can be obtained from the following, which have been excerpted from Table 3.3.3 of the NDS.

Table 10 Effective span length values (NDS 2015)

Load condition	$I_u/d < 7$	$I_u/d < d$
Uniformly distributed load	$I_e = 2.06\, I_u$	$I_e = 1.631\, I_u + 3d$
Concentrated load at center without lateral support at load	$I_e = 1.80\, I_u$	$I_e = 1.631\, I_u + 3d$
Concentrated load at center with lateral support at load	$I_e = (1.11)\, I_u$	$I_e = (1.11)\, I_u$
Two equal loads at 1/3 points with lateral support at loads	$I_e = 1.68\, I_u$	$I_e = 1.68\, I_u$
Three equal loads at 1/4 points with lateral support at loads	$I_e = 1.541\, I_u$	$I_e = 1.541\, I_u$
Four equal loads at 1/5 points with lateral support at loads	$I_e = 1.68\, I_u$	$I_e = 1.68\, I_u$
Equal end moments	$I_e = 1.841\, I_u$	$I_e = 1.841\, I_u$

- Beam stability factor equation:

$$C_L = \left(\left(\frac{1+Q}{1.9}\right) - \sqrt{\left(\frac{1+Q}{1.9}\right)^2 - \frac{Q}{0.95}}\right) \le 1.0$$

$$Q = \frac{F_{be}}{F_b^*}$$

$$F_{be} = \frac{k_{be}E'_y}{R_B^2}$$

$F_b^* = F_b$ for all C's except C_{Fu}, C_V, and C_L.

$$K_{be} = \begin{cases} 0.439 \ VGL \\ 0.561 \ MEL \\ 0.610 \ Glulam \ beams \ and \ MSR \end{cases}$$

VGL: Visually graded lumber

MEL: Machine evaluated lumber

MSR: Machine stress rated lumber

$E'_y = E_y(C_M C_t)$ bending about the y-axis

Table 11 Effective length (NDS Table 3.3.3)

Effective Length (N.D.S. Table 3.3.3.)					
	Loading	Bracing	ℓ_e		
Condition	Condition	Condition	$\ell_u/d < 7$	$7 \leq \ell_u/d \leq 14.3$	$\ell_u/d > 14.3$
	Any condition not listed below		$2.06\ell_u$	$1.63\ell_u + 3d$	$1.84\ell_u$
Single span	Concentrated load at midspan	braced at ends only	$1.80\ell_u$	$1.37\ell_u + 3d$	$1.37\ell_u + 3d$
	uniformly distributed load	braced at ends only	$2.06\ell_u$	$1.63\ell_u + 3d$	$1.63\ell_u + 3d$
Cantilever	concentrated load at unsupported end	-	$1.87\ell_u$	$1.44\ell_u + 3d$	$1.44\ell_u + 3d$
	uniformly distributed load	-	$1.33\ell_u$	$0.90\ell_u + 3d$	$0.90\ell_u + 3d$
				ℓ_e	
Single Span of length L	Uniformly spaced concentrated load	braced at each concentrated load			
	one load	$\ell_u = L/2$		$1.11 \ell_u$	
	two loads	$\ell_u = L/3$		$1.68 \ell_u$	
	three loads	$\ell_u = L/4$		$1.54 \ell_u$	
	four loads	$\ell_u = L/5$		$1.68 \ell_u$	
	five loads	$\ell_u = L/6$		$1.73 \ell_u$	
	six loads	$\ell_u = L/7$		$1.78 \ell_u$	
	seven or more loads			$1.84 \ell_u$	
	Equal end moments	-		$1.84 \ell_u$	

Example 10:

Determine if a Visually Graded Beam 4x8 No.1 DF-L with MC=19% beam is adequate for this load arrangement. Note: The Beam is fully unbraced.

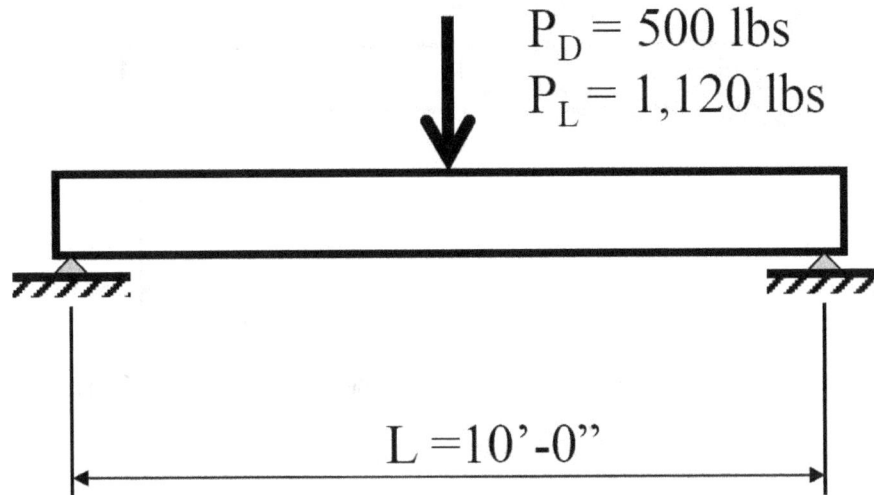

$P_D = 500$ lbs
$P_L = 1,120$ lbs

$L = 10'\text{-}0''$

The general solution procedure is to test if the following is true:

$$F_b \leq F'_b$$

For lateral stability the beam stability Adjustment factor, C_L, needs to be computed. Note that $C_L \leq 1.0$.

1- The steps for getting the actual stress, F_b, include:

   a) Loading diagram
   b) Determine the critical bending moment
   c) Calculate the actual beam stress

   Therefore,

a) The loading diagrams

Beam self-weight of:

$$\rho = 62.4 \left(\frac{0.5}{1 + G(0.009)(19)} \right) \left(1 + \frac{19}{100} \right) = 34.2 \text{ pcf/ft}$$

$$W_{4x8} = 34.2 \left(\frac{(3.5)(7.25)}{144} \right) = 6.03 \, plf$$

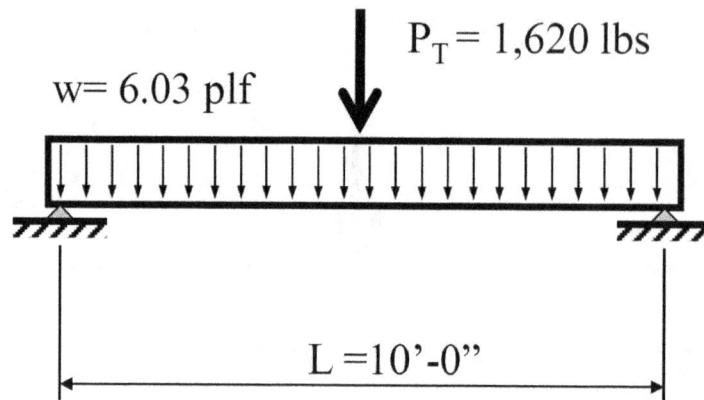

For each of analysis, this load can be considered separately in two parts:

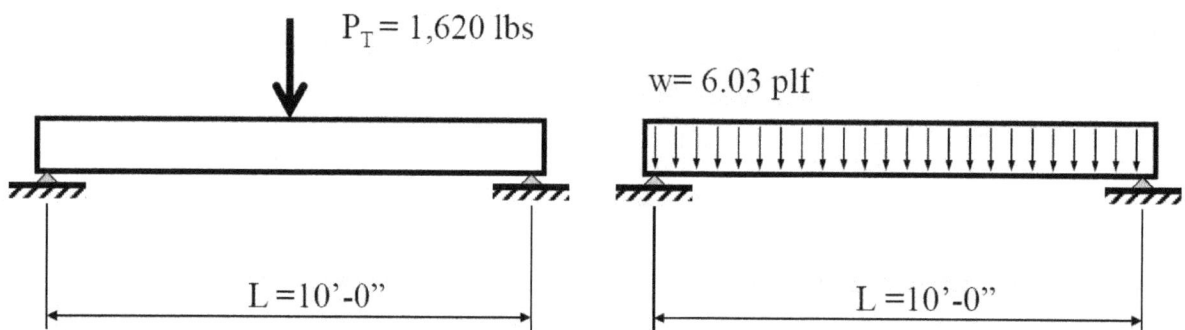

Note: Superposition is valid when the maximum moments for the two different load cases are located at the same place along the beam span, as is true here.

b) Critical bending moment:

Through the superposition of maximum bending moments from a single point load and a uniformly distributed load, the critical moment at mid-span is:

$$M_{max} = \frac{PL}{4} + \frac{WL^2}{8}$$

$$M_{max} = \frac{1,620(10)}{4} + \frac{6.03(10^2)}{8} = 4,125.34 \, ft - lbs$$

c) Actual bending stress:

$$f_b = \frac{M_{max}}{S_x}$$

$$S_x = \frac{bd^2}{6} = \frac{3.5(7.25^2)}{6} = 30.66 \ in^3$$

Then,

$$f_b = \frac{(4{,}125.34)(12)}{30.66} = 1{,}615 \ psi$$

2- The allowable bending stress, F_b':

The allowable bending stress is obtained by adjusting the tabulated base design values by the appropriate adjustment factors as shown below:

$F_b' = F_b \ (C_D \ C_M \ C_t \ C_L \ C_F \ C_{fu} \ C_r)$

F_b = 1000 psi. (NDS Table 4A for a No. 1 DF-L member.)

C_D = 1.25 (The governing load combination consists of dead and roof live load).

C_M = 1.0 (In-service moisture content is less than 19%)

C_t = 1.0 (In-service temperature is within ordinary ranges which is less than 100 degrees F)

C_F = 1.3 (NDS Table 4A)

C_r = 1.0 (member spaces more than 24" o.c.)

C_{fu} = 1.0 (Member is loaded in edge-wise bending)

$F_b' = F_b \ (C_D \ C_M \ C_t \ C_L \ C_F \ C_{fu} \ C_r) = 1{,}625$ psi

Note: Stress on beam is less than F_b'. So, this indicates that beam is adequate to take the imposed load but it may require lateral support. This is checked next.

3- Bending Stability Factor (C_L)

$$K_{be} = 0.439 \ (VGL)$$

$$E'_y = E_y(1)(1) = 1{,}700{,}000 \ psi$$

$$l_u = 10\,ft$$

$$\frac{l_u}{d} = \frac{10(12)}{7.25}\,16.6 > 14.3 \; so \; use \; third \; equation \; to \; compute \; l_e.$$

Since point load condition dominates:

$$l_e = 1.84(10)(12) = 220.8''$$

$$R_B = \sqrt{\frac{220.8(7.25)}{3.5^2}} = 11.4 < 50 \; \textbf{O.K.}$$

$$F_{be} = \frac{0.439(1,700,000)}{11.4^2} = 5,711\,psi$$

$$Q = \frac{5,711}{1,625} = 3.51$$

$$C_L = \left(\left(\frac{1+3.51}{1.9} \right) - \sqrt{ \left(\frac{1+3.51}{1.9} \right)^2 - \frac{3.51}{0.95} } \right) = 0.98 \le 1.0$$

The adjusted allowable stress becomes:

$F_b' = 1000(1.25)(1.3)(0.98) = 1,594$ psi

4- Compare actual stress to allowable stress to determine if the beam is adequate in bending:

$$f_b = 1,615\,psi$$
$$F_b' = 1,594\,psi$$

Since $f_b > F_b'$, this beam is not adequate in bending.

(How to change this design to make this beam work?)

Answer: Provide full blocking or partial blocking at the point load. The designer should check that support at each ¼ length of the beam will work well. $C_L = 0.996$ for ¼ L support.

$F_b' = 1000(1.25)(1.3)(0.996) = 1,618$ psi

Since $f_b < F_b'$, this beam is not adequate in bending provided there are lateral supports at each ¼ L.

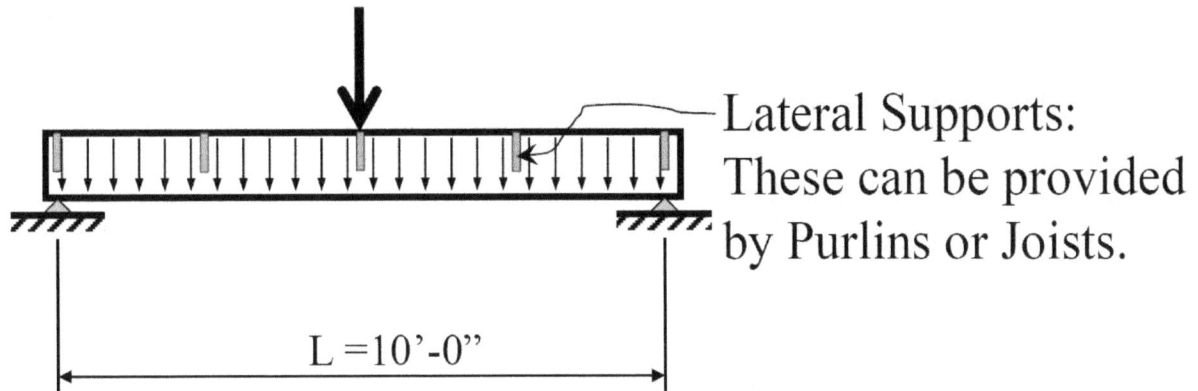

Lateral Supports: These can be provided by Purlins or Joists.

L =10'-0"

2.3 Design of Wood Beams: Shear

Shear in Wood Beams:

Shear normally controls the design of short (L/d ≤ 10), heavily loaded beams.

The n.a. of the beam is the location where shear stresses are maximized. In addition to horizontal shear stresses, vertical shear stresses of the same magnitude simultaneously occur.

$$f_v = \frac{VQ}{Ib}$$

V = shear load as a function of (span, loads, supports).

b = beam width at the point you are investigating.

Q = 1st moment of the area sliding past the point investigating.

I = gross section moment of inertia.

For rectangular beam:

$$f_{vmax} = \frac{3}{2}\left(\frac{V}{A}\right)$$

Use a reduced value for V in fv calculations if:

- In the usual loading case, the beams are fully supported by bearing on one side, and the loads are applied to the other edge.
- This reduced value is the shear at a distance "d" from the face of the support, which is V'.
- For example, consider the shear diagram for a simply supported, uniformly loaded beam.

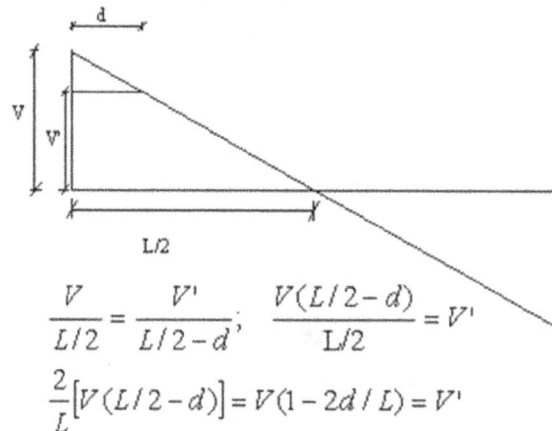

$$\frac{V}{L/2} = \frac{V'}{L/2 - d}; \quad \frac{V(L/2 - d)}{L/2} = V'$$

$$\frac{2}{L}\left[V(L/2 - d)\right] = V(1 - 2d/L) = V'$$

$$V' = V(1 - \frac{2d}{L})$$

A reduced V can be used in F_v because loads are transmitted to support by diagonal compression, not shear, within this "d" zone (i.e. diagonal compression, the critical mode).

- For design:

$$F_v \leq F_v'$$

Where $F_v' = F_v(C_D C_M C_t C_H)$

With F_v = tabulated horizontal shear stress

- The shear strength of wood parallel to grain, (horizontal), is normally much weaker than the shear strength perpendicular to the grain, (vertical).

  C_H = Shear strength factor ≥ 1.0 for solid-sawn.

  Not widely used in design, because it requires knowledge of split lengths and anticipated behavior of splits.

May, however, be useful in evaluating an existing in-service condition. = 1.0 for glulams (recognizing that extensive splitting doesn't occur in a manufactured product).

Example 11:

Determine if the following beam is adequate in shear. Note that the beam has been marked as 24F-V1 SP/SP with MC > 19%.

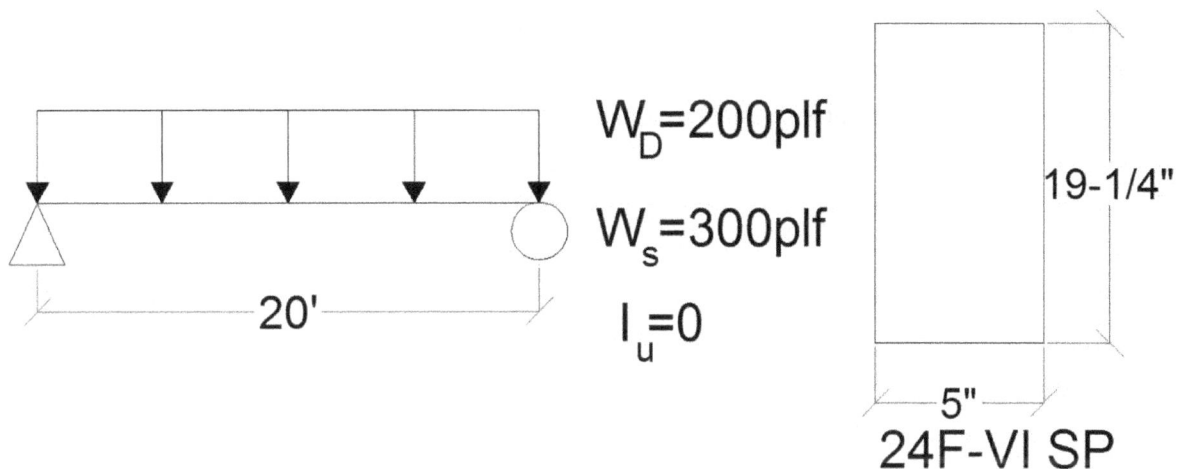

W_D=200plf

W_s=300plf

l_u=0

19-1/4"

5"

24F-VI SP

Check governing load combination:

$$D = \frac{200}{0.9} = 222.22 \; plf$$

$$D + S = \frac{200 + 300}{1.15} = 434.78 \; plf$$

1. Obtain Design Shear V'

$$V = \frac{WL}{2} = \frac{500(20)}{2} = 5{,}000 \; lbs$$

$$V' = V\left(1 - \frac{2d}{L}\right) = 5{,}000\left(1 - \frac{2\left(\frac{19.25}{12}\right)}{20}\right) = 4{,}198 \; lbs$$

Obtain maximum shear stress in the beam

$$f_{vmax} = \frac{3}{2}\left(\frac{V'}{A}\right) = \frac{3}{2}\left(\frac{4,198}{25(19.25)}\right) = 65.4\ psi$$

Obtain allowable shear stress in the beam

Where $F_v' = F_v(C_D C_M C_t C_H)$

$$F_v = 300\ psi$$

$$C_t = 1.0$$

$$C_H = 1.0$$

Note that for Glu-Lam dry is defined at an MC=16%

$$C_M = 0.875$$

Table 12 Wet service factor for Glued Laminated beam (NDS 2015)

WET SERVICES FACTORS, C_M					
When structural glued laminated timber is used where moisture content will exceed 16% or greater, design values shall be multiplied by appropriate wet service factor form the following table:					
F_b	F_t	F_v	$F_{c\perp}$	F_c	E
0.8	0.8	0.875	0.53	0.73	0.833

$$C_D = 1.15$$

$$F_v' = 300\big((1.15)(0.875)(1.0)(1.0)\big) = 302\ psi$$

Compare actual stress to allowable stress:

$$f_{vmax} = 65.4\ psi$$

$$F_v' = 302\ psi$$

Answer: Since $f_v < F_v'$, this beam is adequate in shear.

Table 5 of NDS (NDS 2015) indicates the design valued for structural Glued laminated timber.

2.4 Design of Wood Beams (Deflection)

- IBC02, Table 16-D for the roof and floor member supporting plaster ceilings.

- For load combinations, limitations see page C.3 for Table 1604.3

- AITC recommendations are as follows:

    Note: Applied load in these tables refer to live load.

Table 13 Recommended deflection limits (IBC 2015)

Recommended Deflection Limitations		
Use Classification	Applied load only	Applied load + dead load
Roof beams		
Indestrial	L/180	L/120
Commercia and institutional		
Without plaster ceiling	L/240	L/180
With plaster ceiling	L/360	L/240
Floor beams		
Ordinary usage*	L/360	L/240
Highway bridge stringers	L/300	
Railway bridge stringers	L/300 to L/400	

"The ordinary usage classification is for floore intended for construction in which walking comfort and minimized plaster cracking are the main consideration. These recommended deflection limits may not eliminate all objections to vibrations such as in long spans approaching the maximum limits or for some office and institutional applications where incrreased floor stiffness is desired. For this usage the deflection limitations in th following table have been found to provide additional stiffness.

Deflection Limitations for Uses Where Increeased Floor Stiffness Is Desired		
Use Classification	Applied load only	Applied load + K (dead load)*
Floor beams		
Commercial, Office and institutional		
Floor hoints, spams to 26 ft†		
L ≤ 60 psf	L / 480	L / 360
60 psf ≤ L ≤ 80 psf	L / 480	L / 360
L ≥ 80 psf	L / 420	L / 300
Girders, spans to 36 ft†		
L ≤ 60 psf	L / 480‡	L / 360
60 psf ≤ L ≤ 80 psf	L / 420‡	L / 300
L ≥ 80 psf	L / 360‡	L / 240

*K = 1.0 except for seasoned members where K = 0.5. Seasoned members for this usage are defined as having a moisture content of less than 16 percent at the time of installation.
†For gider spans greater than 36 ft and joints spans greater than 26 ft, special design consideations may be required such as more restrictive deflection limits and vibration consideration that include the total mass of the floor.
++Based on reduction of live load as permitted by the Code.

Table 14 Values of K for different materials (IBC 2015)

WOOD		*REINFORCED CONCRETE*[2]	STEEL
unseasoned	*seasoned*[1]		
1.0	0.5	$T/(1+50p')$	0

1- Seasoned lumber is lumber having a moisture content of less than 16 percent at time of installation and used under dry condition of use such as in covered stractuers.

2- See also section 1909 for definition and other requirements.

p′ shall be the value at midspan for simple and conditions spans, and a support for contilevers. Time-dependent factor T for sustained loads may be taken equal to:

Five years or more	2.0
twelve months	1.2
six months	1.4
three months	1.0

Also, a minimum roof slope of 1/4"/ft for all roof members to prevent ponding. Rule of thumb: any roof member should be stiff enough so that a 5 psf uniform load results in $\Delta \leq 1/2$" to minimize ponding potential.

The governing equation for deflection is $\Delta_{max} \leq \Delta_{all}$.

Δ_{all} criteria are often given in terms of the member's span, L, in inches.

Δ_{max}= Function of (P, W, L, E′, and I) where $E' = E(C_M C_t)$.

E = tabulated modulus of elasticity about the axis of bending - given as an average value.

C_t = buckling stiffness factor used primarily in and therefore takes a value of 1 for beams.

Note: C_D does not apply to E.

Camber = initial curvature built into the member to oppose the deflection of the beam under gravity loads.

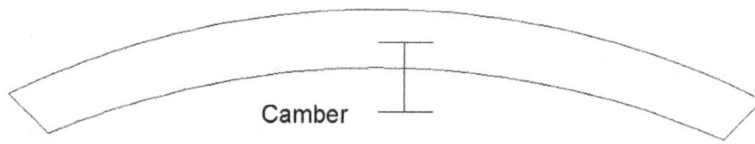

Camber

- Typical glulam camber = 1.5 Δ_D.
- Member produces a nearly level beam under long-term deflection.
- Creep = the time-dependent deformation that develops at a slow, but steady rate over long periods of time.
- Creep is greater for wood members drying under load or exposed to varying temperature and humidity conditions.
- To estimate the long-term deflection accounting for creep, increase the long-term load component by 1.5:
- Long Term, Total Δ estimates for seasoned lumber or glulams:

$$\Delta_{total} = \underbrace{(1.5\Delta_{DL+\%LL})}_{\text{Long -Term}} + \underbrace{\Delta_{LL}}_{\text{Short-Term}}$$

Example 12:

Check deflection limit for example 8.

Note from example 8: USE 4 x 8 DF-L No. 1

Step 1: Deflection under Live Load

E= 1,700,000 psi (Use this because code allows the average E)

$C_M = 1.0$; $C_T = 1.0$

$E' = E(1)(1) = 1,700,000\ psi$

$$\Delta_L = \frac{PL^3}{48E'I_{xx}} = \frac{1,000(8 \times 12)^3}{48(1,700,000)(111.1)} = 0.098 \; in.$$

$$\Delta_{all} = \frac{L}{360} = \frac{(8 \times 12)}{360} = 0.27 \; in. \geq 0.098 \; in. \therefore \text{ Design is okay for live load}$$

Step 2: Deflection under Dead + Live Load

$$\Delta_{D+L} = \frac{5D_B L^4}{384E'I_{xx}} + \frac{(D+L)(L)^3}{48 \times E'I_{xx}}$$

$$\Delta_{D+L} = \frac{5(6.168)(8 \times 12)^4}{384 \times 1,700,000 \times 111.1} + \frac{(1,000 + 1,000)(8 \times 12)^3}{48 \times 1,700,000 \times 111.1}$$

$$\Delta_{D+L} = 0.036 + 0.098 = 0.232 \; in.$$

$$\Delta_{all} = \frac{L}{240} = \frac{8 \times 12}{240} = 0.40 \; in. \geq 0.232 \; in.$$

$$\therefore \text{ Design is okay for Dead + Live Load}$$

Design of Wood Beams (Bearing Stresses)

Bearing stresses perpendicular to the grain beams bearing directly on:

Another wood member, Masonry wall, Steel Hanger, Bearing plates, and Column Cap.

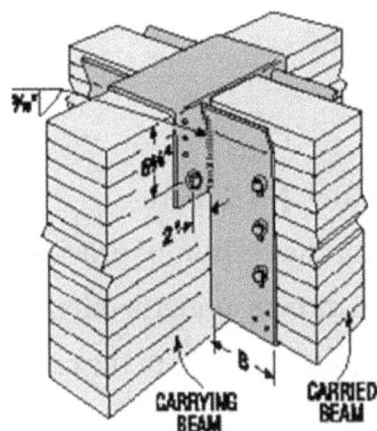

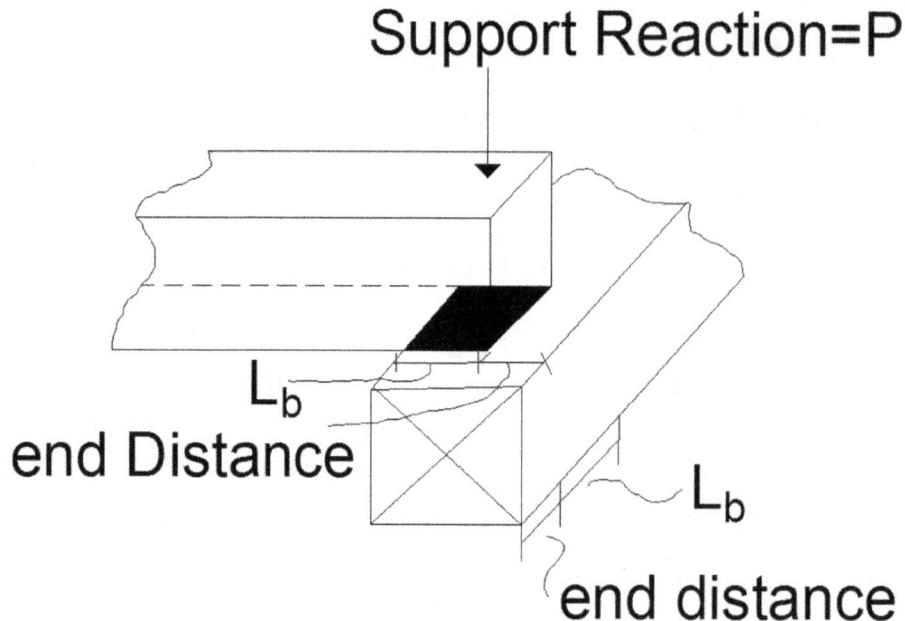

Support Reaction=P

L_b

end Distance

L_b

end distance

Bearing stresses occur where there is a transfer of load from the beam to its supporting member. Bearing failures are due to the localized crushing of wood fiber in the vicinity of the bearing load.

- These failures are not serious and do not lead to structural collapse.
- The allowable tabulated compression perpendicular to the grain values upon an average deformation level of 0.04".

- The governing equation is:

$$f_{c\perp} \leq F'_{c\perp}$$
$$f_{c\perp} = \frac{P}{A}$$

A is the bearing Area (bl_b)

$$F'_{c\perp} = F_{c\perp}(C_M C_t C_b) \text{ Note: No } C_d \text{ here}$$

C_b is the bearing area factor. It is accounted for the resistance provided by surrounding wood upon deformation of wood directly underneath the load. At locations where $l_b < 6''$ and distance from the end of beam to the contact area is more than $3''$, then

$$C_b = 1 + \frac{0.375}{l_b}$$

In the tabular form, C_b for various bearing length is:

Table 15 bearing area factor for various bearing length

ℓ_b	0.5"	1"	1.5"	2"	3"	4"	≥ 6"
C_b	1.75	1.38	1.25	1.19	1.13	1.10	1.00

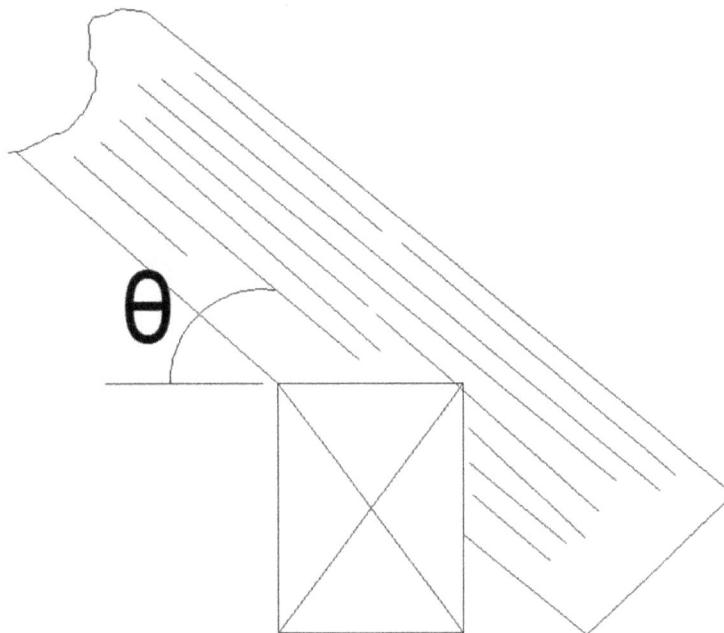

- For bearing situations that occur somewhere between perpendicular or parallel to the grain use Hankinson's Formula to determine the allowable stress value.

$$F_\theta' = \frac{F_{C\parallel}' F_{C\perp}'}{F_{C\parallel}' \sin^2 \theta \, F_{C\perp}' \cos^2 \theta}$$

Arzhang Zamani

Example 13:

To prevent bearing failure, what length of connection bracket is required?

Note that L=12' is the distance from the face of the support and not the centerline.

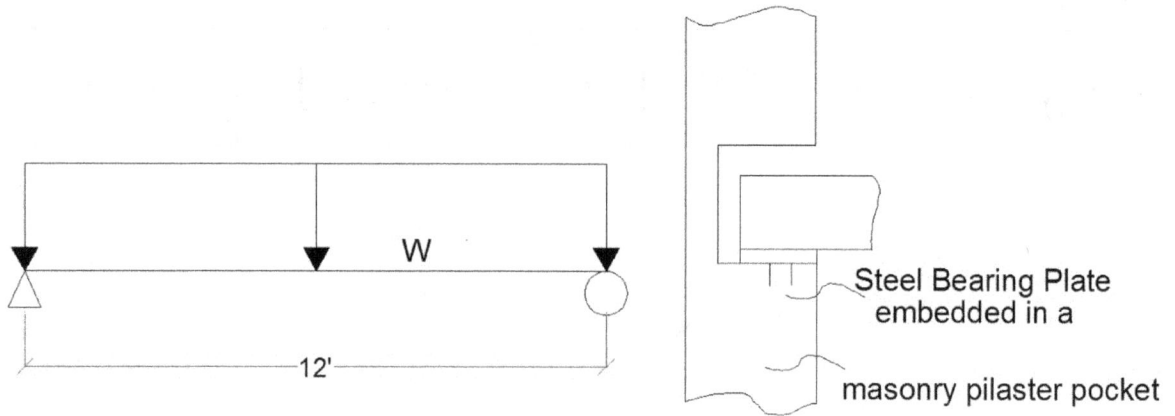

These dry floor beams are No. 1 and better, DF-L, 4X16 joists spaced at 4' o.c., This is a heavy storage facility. Loads:

Dead: beam dead weight + decking + misc. Total D = 130 plf

Table 16 Dead load breakdown

Assume 35 pcf wood:	13 plf
Assume 3/4" plywood subfloor with 2" reinforced concrete topping:	3(3/4)(4) + 150(2/12)(4) = 109 plf
Misc. at 2 psf:	2(4) = 8 plf
Total D	130 plf

Floor live load: heavy storage = 250 psf

L = 250(4) = 1,000 plf

Therefore, $W_{tl} = 130 + 1,000 = 1,130 \, plf$

Examine:

$$f_{c\perp} \leq F'_{c\perp}$$

$$f_{c\perp} = \frac{P}{A} = \frac{\frac{WL}{2}}{3.5l_b} = \frac{\frac{(1,130)(12)}{2}}{3.5l_b} = \frac{6,780}{3.5l_b}$$

A is the bearing Area

$$F'_{c\perp} = F_{c\perp}(C_M C_t C_b C_i)$$
$$F_{c\perp} = 625 \, psi$$

Assume for now $C_b = 1$

$\frac{6,780}{3.5l_b} = 625$ Solve for l_b: $l_b = 3.1 \, in.$

Try $l_b = 3.1 \, in.$ Then, $C_b = 1.13$

$$F'_{c\perp} = F_{c\perp}(C_M C_t C_b) = 625(1.13) = 706 \, psi$$

$$L_t = 12 + \frac{3}{12} = 12.25 \, ft$$

$$f_{c\perp} = \frac{P}{A} = \frac{\frac{WL}{2}}{3.5l_b} = \frac{\frac{(1,130)(12.25)}{2}}{3.5(3)} = 659 \, psi$$

Since $f_{c\perp} \leq F'_{c\perp}$, this beam is adequate for bearing stress.

Example 14:

A simply supported roof wood beam spans 12 feet, and in addition to its self-weight, the beam carries a uniform dead load of 100 pounds per linear foot plus a uniform roof live load of 250 pounds per linear foot.

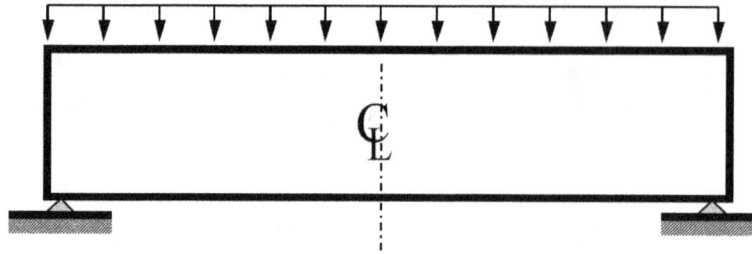

If the beam is a Douglas Fir-larch, No2, with a moisture content of 18%, what size beam should be used?

Design/check the beam for the following conditions:

Bending (the beam is fully supported), Shear, Deflection, Bearing

Example 14 Solution:

Simply supported beam - Distributed Load

$$DF - L \ \& \ mc = 18 \ \& \ G = 0.5$$

$$\text{No.2} \ F_b = 900 \ psi \ \ L = 12 \ ft$$

$$\rho = 62.4 \left(\frac{G}{1 + G(0.009)(mc)} \right) \left(1 + \frac{mc}{100} \right) = 34.057 \ pcf$$

$$D = 100 \ plf \quad L_1 = 250 \ plf$$

Try 1:

Flexure:

$$F'_b = F_b (C_D C_M C_t C_L C_F C_{fu} C_r)$$

Obtaining governing load combination:

$CD_1 = 0.9$

$$\frac{D}{C_{D1}} = 111.111 \ plf$$

$C_{D2} = 1.0$

$\frac{D+L_1}{C_{D2}} = 350\ plf$ Governs

So, $C_F = 1$

$C_M = 1\ (mc < 19\%)$

$C_t = 1$

$C_L = 1$ (To be checked)

$C_r = 1$

$C_{fu} = 1$ (Loaded in strong axis)

$$F'_b = F_b\left(C_D C_M C_t C_L C_F C_{fu} C_r\right) = 900\ psi$$

$$W = D + L_1 = 350\ plf$$

$$M_{max} = W\frac{l^2}{8} = 75{,}600\ lbf - in$$

$$S_{req} = \frac{M_{max}}{F'_b} = 84\ in^3$$

Use 4x14 $C_F = 1$ & $b = 3.5\ in.$ & $d = 13.25\ in.$

$$S_1 = \frac{bd^2}{6} = 102.411\ in^3$$

$$A_1 = b(d) = 46.375\ in^2$$

$$D_1 = \rho A_1 = 10.968\ plf$$

$$W = D + L_1 + D_1 = 360.97\ plf$$

$$M_{max} = W\left(\frac{l^2}{8}\right) = 77{,}969.1\ lbf.in$$

$$S_{req} = \frac{M_{max}}{F'_b} = 86.632\ in^3$$

Ok, Provided S is greater than required.

Arzhang Zamani

Note: Since the ratio of beam size is 1:4, At the end of members by blocking lateral support is provided. Therefore, the flexural capacity of this beam is appropriate for this demand. The l/d is about 12 which means that most probably bending is governing for this case.

Shear:

Check $C_H = 1$

$$F'_v = F_v(C_D C_M C_t C_H)$$

$$F'_v = F_v(C_D C_M C_t C_H) = 95 \ psi \ \text{(capacity)}$$

$$W = 360.97 \ plf \ L = 12 \ ft$$

$$V = W.\frac{l}{2} = 2{,}165.81 \ lbf$$

$$V' = V(1 - \frac{2d}{l}) = 1{,}767.24 \ lbf$$

$$f_{v.max} = 1.5 \left(\frac{V'}{A_1}\right) = 57.161 \ psi \ \text{OK}$$

Deflection:

$$E = 1{,}600{,}000 \ psi$$

$$E' = E C_M C_t = 1{,}600{,}000 \ psi$$

$$I_{xx} = \frac{1}{12}(b)(d^3) = 678.476 \ in^4$$

Deflection under Live Load:

$$L_1 = 250 \ plf$$

$$\Delta_L = \frac{5(L_1)l^4}{384E'I_{xx}} = 0.107 \ in.$$

$$\Delta_{allowable.Live} = \frac{l}{360} = 0.4 \ in. \text{OK}$$

Note: assumed plaster ceiling

60

Deflection under Dead and Live Load:

$$W = D + L_1 + D_1 = 360.97 \; plf$$

$$\Delta_{D+L} = \frac{5(W)l^4}{384E'I_{xx}} = 0.155 \; in$$

$$\Delta_{allowable.Dead+Live} = \frac{l}{240} = 0.6 \; in. \; OK$$

Note: assumed plaster ceiling

Therefore, the deflection is within the limit.

Bearing:

Goal: $f_{c.prependicular} < F'_{c.prependeicular}$

$$P = \frac{W.l}{2} = 2{,}165.81 \; lbf$$

$$A_b = b.l_b$$

$$f_{c.prependicular} = \frac{P}{A_b}$$

$$F_{c.prependeicular} = 625 \; psi$$

$$F'_{c.prependeicular} = F_{c.prependeicular}(C_M C_t C_b) \; (\text{Capacity})$$

$$C_b = 1 + \frac{0.375}{l_b}$$

$$l_b = \frac{P}{b.F'_{c.prependeicular}}$$

Use: $l_b = 1 \; in.$

$$C_b = + \frac{0.375}{l_b} = 1.375 \; in$$

$$F'_{c.prependeicular} = F_{c.prependeicular}(C_M C_t C_b) = 859.37 \; psi \; (\text{Capacity})$$

$$\frac{P}{A_b} = 618.8 \; psi \; (\text{Demand}) \; OK$$

Arzhang Zamani

Example 15:

Design a Structural Glu-Laminated Wood beam for the given conditions

- Use a 24F-V4 DF/DF, western species
- L=24 ft
- $L_u = 0$ ft
- $MC < 16\%$
- Beam supports a non-plaster ceiling
- The Superimposed Dead Load is $W_D = 150\ plf$ (Note must add the beam self-weight)
- The roof live load $W_L = 250\ plf$ use a $C_D = 1.0$
- The roof snow load $W_L = 350\ plf$ use a $C_D = 1.15$

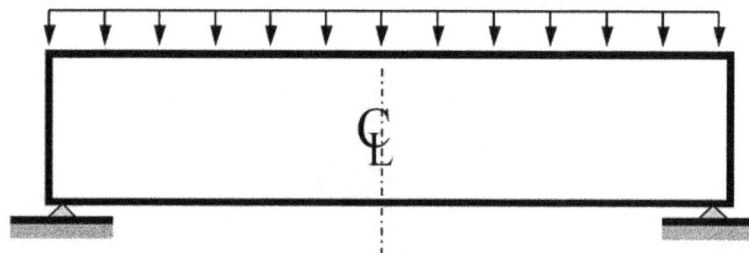

Design/check the beam for the following conditions:

Bending (the beam is fully supported), Shear, Deflection, Bearing

Example 15 Solution:

Simply supported beam - Distributed Load - Glu-laminated beam

$$DF - L\ \&\ mc = 15\ (assumed)\ \&\ G = 0.5$$

24F-V4 $F_b = 2,400\ psi\ L = 24\ ft$

$$\rho = 62.4 \left(\frac{G}{1 + G(0.009)(mc)} \right) \left(1 + \frac{mc}{100} \right) = 33.61\ pcf$$

$$D = 150\ plf$$

$$L_1 = 250\ plf$$

$$S_{snow} = 350 \; plf$$

Try 1:

Flexure:

$$F'_b = F_b(C_D C_M C_t C_L C_F C_{fu} C_r)$$

Obtaining governing load combination:

$CD_1 = 0.9$

$$\frac{D}{C_{D1}} = 166.667 \; plf$$

$C_{D2} = 1.0$

$$\frac{D + L_1}{C_{D2}} = 400 \; plf$$

$C_{D3} = 1.15$

$$\frac{D + S_{snow}}{C_{D3}} = 434.78 \; plf \quad \text{Governs}$$

So, $C_D = 1.15$

Note: For Glu-Laminated Beam the C_V is replaced by C_f.

Assume $C_V = 1$ (To be checked)

$C_M = 1 \; (mc < 16\%)$

$C_t = 1$

$C_L = 1$ (To be checked)

$C_r = 1$

$C_{fu} = 1$ (Loaded in strong axis)

$$F'_b = F_b\left(C_D C_M C_t C_L C_V C_{fu} C_r\right) = 2{,}760 \; psi$$

$$W = D + S_{snow} = 500 \; plf$$

Arzhang Zamani

$$M_{max} = W\frac{l^2}{8} = 432{,}000 \; lbf - in$$

$$S_{req} = \frac{M_{max}}{F'_b} = 84 \; in^3$$

Use 5.125x15 $b = 5.125 \; in. \; \& \; d = 15 \; in.$

$$S_1 = \frac{bd^2}{6} = 192.188 \; in^3$$

$$F'_b = 2{,}760 \; psi$$

$$C_V = 1$$

$$A_1 = b(d) = 78.875 \; in^2$$

$$D_1 = \rho A_1 = 17.94 \; plf$$

$$W = D + D_1 + S_{snow} = 517.94 \; plf$$

$$M_{max} = W\left(\frac{l^2}{8}\right) = 447{,}503.18 \; lbf.in$$

Calculating C_V:

Western Species, x=10, y=10, and z=10 − k=1 (Distributed load)

$$C_V = k\left(\frac{12}{d}\right)^{\frac{1}{x}}\left(\frac{5.125}{b}\right)^{\frac{1}{y}}\left(\frac{21}{l}\right)^{\frac{1}{z}} = 0.965$$

$$F'_b = F_b\left(C_D C_M C_t C_L C_V C_{fu} C_r\right) = 2{,}663.29 \; psi$$

$$S_{req} = \frac{M_{max}}{F'_b} = 168.03 \; in^3$$

Ok, Provided S is greater than required.

Therefore, the flexural capacity of this beam is appropriate for this demand.

Shear:

Check $C_H = 1$

64

$$F_v = 265 \, psi$$

$$F'_v = F_v(C_D C_M C_t C_H)$$

$$F'_v = F_v(C_D C_M C_t C_H) = 304.75 \, psi \text{ (capacity)}$$

$$W = 517.94 \, plf \quad L = 12 \, ft$$

$$V = W \cdot \frac{l}{2} = 6{,}215.32 \, lbf$$

$$V' = V(1 - \frac{2d}{l}) = 5{,}567.89 \, lbf$$

$$f_{v.max} = 1.5 \left(\frac{V'}{A_1}\right) = 108.64 \, psi \text{ OK}$$

Deflection:

$$E = 1{,}800{,}000 \, psi$$

$$E' = E C_M C_t = 1{,}800{,}000 \, psi$$

$$I_{xx} = \frac{1}{12}(b)(d^3) = 1{,}441.4 \, in^4$$

Deflection under Live Load:

$$L_1 = 250 \, plf$$

$$\Delta_L = \frac{5(L_1)l^4}{384 E' I_{xx}} = 0.719 \, in.$$

$$\Delta_{allowable.Live} = \frac{l}{240} = 1.2 \, in. \text{OK}$$

Note: assumed non-plaster ceiling

Deflection under Dead and Live Load:

$$W = D + D_1 + S_{snow} = 517.94 \, plf$$

$$\Delta_{D+L} = \frac{5(W)l^4}{384 E' I_{xx}} = 1.49 \, in$$

Arzhang Zamani

$$\Delta_{allowable.Dead+Live} = \frac{l}{180} = 1.6 \; in. \; \text{OK}$$

Note: assumed non-plaster ceiling

Therefore, the deflection is within the limit.

Bearing:

Goal: $f_{c.prependicular} < F'_{c.prependeicular}$

$$P = \frac{W.l}{2} = 6{,}215.32 \; lbf$$

$$A_b = b.l_b$$

$$f_{c.prependicular} = \frac{P}{A_b}$$

$F_{c.prependeicular} = 650 \; psi$

$F'_{c.prependeicular} = F_{c.prependeicular}(C_M C_t C_b)$ (Capacity)

$$C_b = 1 + \frac{0.375}{l_b}$$

$$l_b = \frac{P}{b.F'_{c.prependeicular}} = 1.866 \; in.$$

Use: $l_b = 2 \; in.$

$$C_b = 1 + \frac{0.375}{l_b} = 1..188 \; in.$$

$F'_{c.prependeicular} = F_{c.prependeicular}(C_M C_t C_b) = 771.875 \; psi$ (Capacity)

$\frac{P}{A_b} = 606.373$ (Demand) OK

Therefore, Use Glu-Laminated beam - 5.125x15   24F-V4 DF/DF   Western Species

2.5 Cantilever Beam System

Consider the three-span beam shown below. In this system $L_1 + l_1$ is the length of the first span and $L_2 + l_2$ is the length of the third span, while the length of the intermediate span or the cantilever system is L.

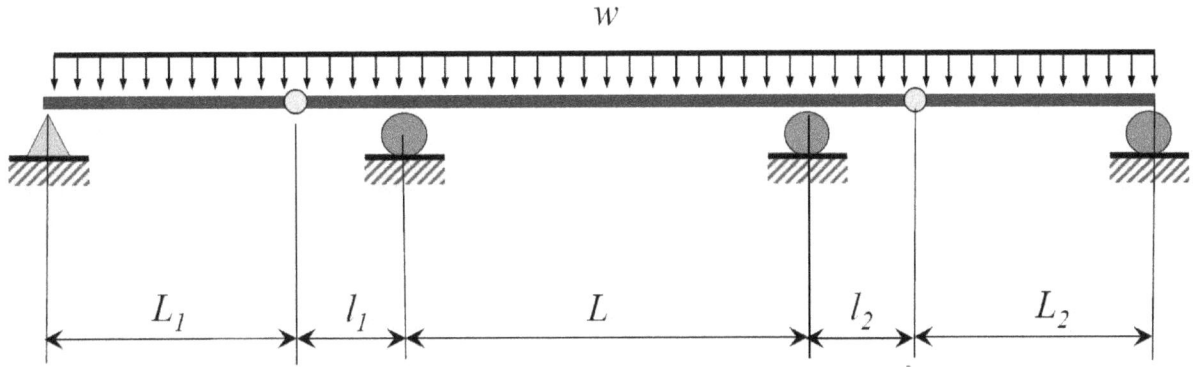

Design of beams within L_1 and L_2 follows the principles discussed in precious sections related design of simply supported beams with uniform loads. The intermediate span is the cantilever system and can be isolated according to the system shown below. In this system, point loads P_1 and P_2 resulted from reactions of the adjacent simply support (or also designated as dropped spans/beams) beams.

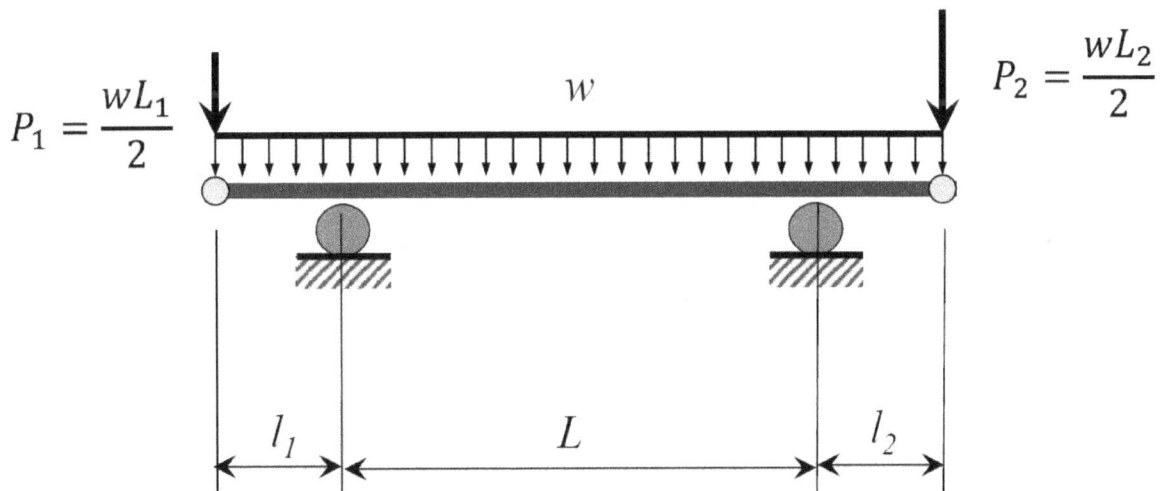

SHEAR FORCE DIAGRAM

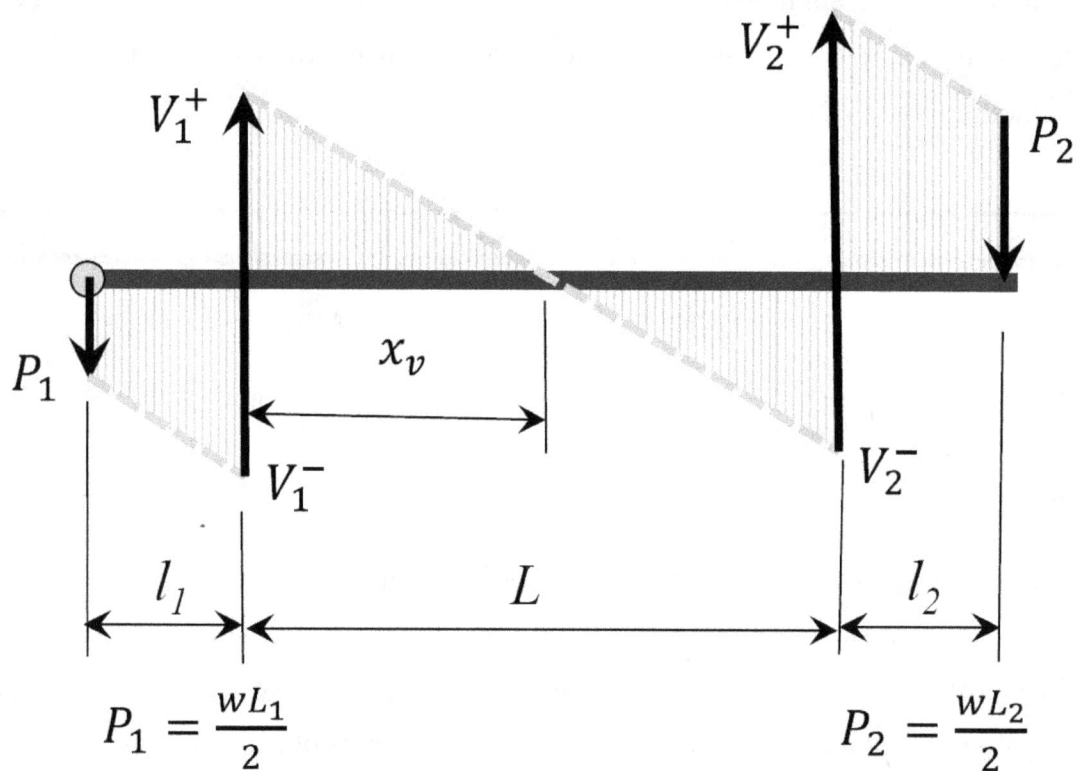

$$P_1 = \frac{wL_1}{2} \qquad\qquad P_2 = \frac{wL_2}{2}$$

$P_1 = \frac{wL_1}{2}$

$V_{x=0} = -P_1$

$V_1^- = -P_1 - wl_1$

$V_2^- = V_1^+ - wL$

$R_2 = \dfrac{-\left[P_1 l_1 - wL_t\left(\frac{L_t}{2} - l_1\right) - P_2(L_t - l_1)\right]}{L}$

$R_1 = wL_t - R_2$

$P_2 = \frac{wL_2}{2}$

$V_{x=L_t} = +P_2$

$V_1^+ = V_1^- + R_1$

$V_2^+ = V_2^- + R_2$

$x_v = \dfrac{|V_1^+|L}{\left(|V_1^+| + |V_2^-|\right)}$

BENDING MOMENT DIAGRAM

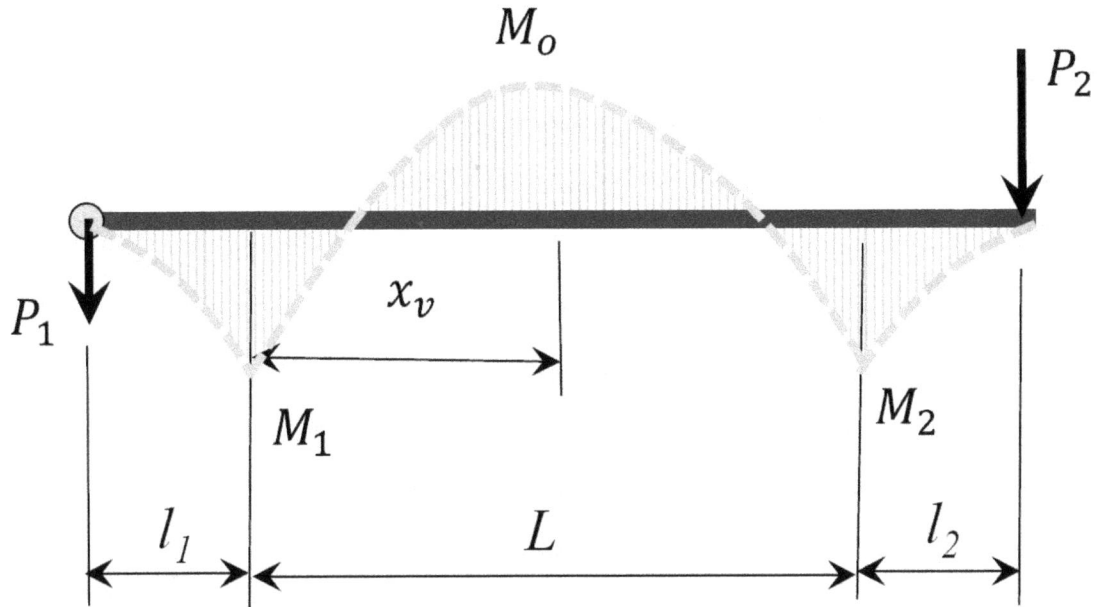

$$M_1 = \frac{(V_{x=0} + V_1^-)}{2} l_1$$

$$M_0 = M_1 + \frac{V_1^+ x_v}{2}$$

$$M_2 = M_0 + \frac{V_2^-(L - x_v)}{2}$$

END MOMENT DEFLECTION AT MIDSPAN

$$d_{x,M_1} = \frac{0.0542 M_1 L^2}{EI}$$

$$d_{x,M_2} = \frac{0.0542 M_2 L^2}{EI}$$

A design example of cantilever beam system.

Example 16:

Design a Structural Glu-Laminated Wood beam for the given conditions:

1. Use a 24F-V4 DF/DF, western species

2. L_u design bracing as required.

3. $M_C < 16\%$.

4. Beam supports the non-plaster ceiling.

5. $W_D = 200\text{plf}$.

6. $W_{Lr} = 320\text{plf}$.

7. $W_S = 300\text{plf}$.

8. Must consider unbalanced loading

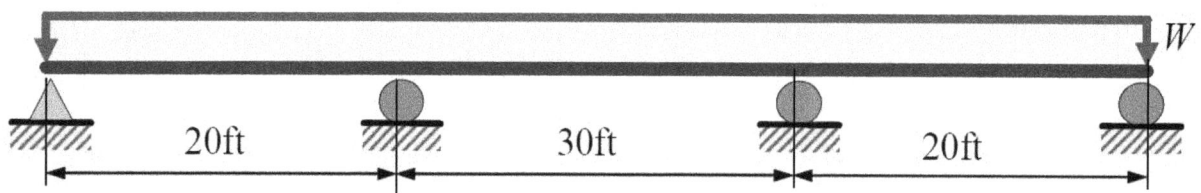

Step 1. Locate Hinges, optimizing for length:

a. Case 1. Hinges on the end spans

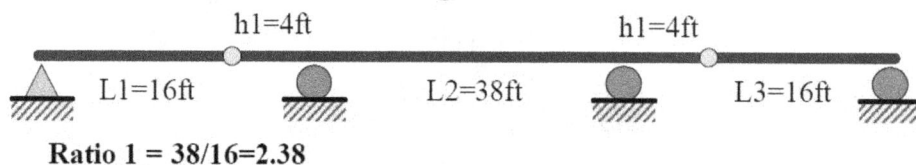

Ratio 1 = 38/16=2.38

b. Case 2. Hinges on the center span

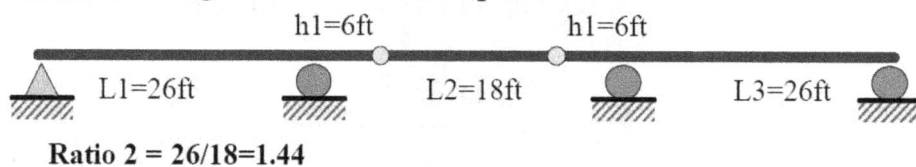

Ratio 2 = 26/18=1.44

Question:

Which one of these will give a better design alternative?

As can be seen, both beams are statically determinant. Therefore, designing each case is straightforward. The better design alternative is the design which has the minimum required volume of wood. A sample of design is presented in the design project in the appendix.

Sample of tables extracted from Simpson Strong-Tie Catalog (Simpson 2017) are shown on the next pages

These products are available with additional corrosion protection. Additional products on this page may also be available with this option, check with Simpson Strong-Tie for details.

Model No. (CC shown ECC/ECCU similar)	Beam Width	Dimensions						Machine Bolts					Allowable Loads (DF/SP)				Code Ref.	CCO Model No. (No Legs)	ECCO Model No. (No Legs)
		W1	W2	L			H1	Size	Beam			Post	Down		Uplift				
				CC	ECC	ECCU			CC	ECC	ECCU		CC	ECC/ ECCU	CC (160)	ECCU (160)			
CC3¼-4	3⅛	3¼	3⅝	11	7½	9½	6½	⅝	4	2	4	2	16980	6125	3640	1010	I12, L4, F11	CCO3¼	ECCO3¼
CC3¼-6	3⅛	3¼	5½	11	7½	9½	6½	⅝	4	2	4	2	19250	9625	3640	1010			
CC44	4x	3⅝	3⅝	7	5½	6½	4	⅝	2	1	2	2	15310	7655	1465	205		CCO4	ECCO4
CC46	4x	3⅝	5½	11	8½	9½	6½	⅝	4	2	4	2	24060	12030	2800	740		CCO4/6	ECCO4/6
CC48	4x	3⅝	7½	11	8½	9½	6½	⅝	4	2	4	2	24060	16405	2800	740	160		
CC4.62-3.62	4½	4⅝	3⅝	11	8½	9½	6½	⅝	4	2	4	2	19020	7655	2800	740	170	CCO4.62	ECCO4.62
CC4.62-4.62	4½	4⅝	4⅝	11	8½	9½	6½	⅝	4	2	4	2	24450	9845	2800	740			
CC4.62-5.50	4½	4⅝	5½	11	8½	9½	6½	⅝	4	2	4	2	28585	12030	2800	740			
CC5¼-4	5⅛	5¼	3⅝	13	9½	10½	8	¾	4	2	4	2	26635	10045	7530	2735	I12, L4, F11	CCO5¼	ECCO5¼
CC5¼-6	5⅛	5¼	5½	13	9½	10½	8	¾	4	2	4	2	28190	15785	7530	2735			
CC5¼-8	5⅛	5¼	7½	13	9½	10½	8	¾	4	2	4	2	37310	21525	7530	2735			
CC64	6x	5½	3⅝	11	7½	9½	6½	⅝	4	2	4	2	28586	12030	4040	1165		CCO6	ECCO6
CC66	6x	5½	5½	11	7½	9½	6½	⅝	4	2	4	2	30250	18905	4040	1165			
CC68	6x	5½	7½	11	9½	9½	6½	⅝	4	2	4	2	37810	25780	4040	1165			ECCO68
CC6-7⅛	6x	5½	7⅛	11	9½	9½	6½	⅝	4	2	4	2	37810	24060	4040	1165	160		
CC74	6¾	6⅞	3⅝	13	10½	10½	8	¾	4	2	4	2	33490	13230	7525	3605	170	CCO7	ECCO7
CC76	6¾	6⅞	5½	13	10½	10½	8	¾	4	2	4	2	37125	20790	7525	3605	I12, L4, F11		
CC77	6¾	6⅞	6⅞	13	10½	10½	8	¾	4	2	4	2	49140	25515	7525	3605			
CC78	6¾	6⅞	7½	13	10½	10½	8	¾	4	2	4	2	49140	28350	7525	3605			
CC7⅛-4	7	7⅛	3⅝	13	10½	10½	8	¾	4	2	4	2	34736	18375	7510	4855	160	CCO7⅛	ECCO7⅛
CC7⅛-6	7	7⅛	5½	13	10½	10½	8	¾	4	2	4	2	58500	28875	7585	4855			
CC7⅛-7⅛	7	7⅛	7⅛	13	10½	10½	8	¾	4	2	4	2	57750	36750	7585	4855			
CC7⅛-8	7	7⅛	7½	13	10½	10½	8	¾	4	2	4	2	52500	36750	7585	4855			
CC84	8x	7½	3⅝	13	10½	10½	8	¾	4	2	4	2	37210	16405	7440	2625	I12, L4, F11	CCO8	ECCO8
CC86	8x	7½	5½	13	10½	10½	8	¾	4	2	4	2	41250	23100	7440	2625			
CC88	8x	7½	7½	13	10½	10½	8	¾	4	2	4	2	54600	31500	7440	2625			
CC94	8¾	8⅞	3⅝	13	10½	10½	8	¾	4	4	4	2	43410	19905	7515	3990		CCO9	ECCO9
CC96	8¾	8⅞	5½	13	10½	10½	8	¾	4	4	4	2	48125	26950	7515	4670			
CC98	8¾	8⅞	7½	13	10½	10½	8	¾	4	4	4	2	63700	36750	7515	4670			
CC106	10x	9½	5½	13	10½	10½	8	¾	4	4	4	2	52250	29260	7515	3325		CCO10	ECCO10

TOP FLANGE HANGERS LEG/MEG/EG *Beam & Glulam Hangers*

See Hanger Options on pages 233-243 for hanger modifications, which may result in reduced loads.

Designed to support large members typically found in glulam beam construction.

MATERIAL: See table

FINISH: Simpson Strong-Tie gray paint. Hot-dip galvanized available; specify HDG.

INSTALLATION: • Use all specified fasteners. See General Notes.
- Maintain minimum 4D end distance and edge distance from bolt to end of header and nearest loaded edge per NDS requirements.

OPTIONS: • See Hanger Options, pages 233-243.
- Models available without top flanges; see table loads.

CODES: See page 12 for Code Reference Key Chart.

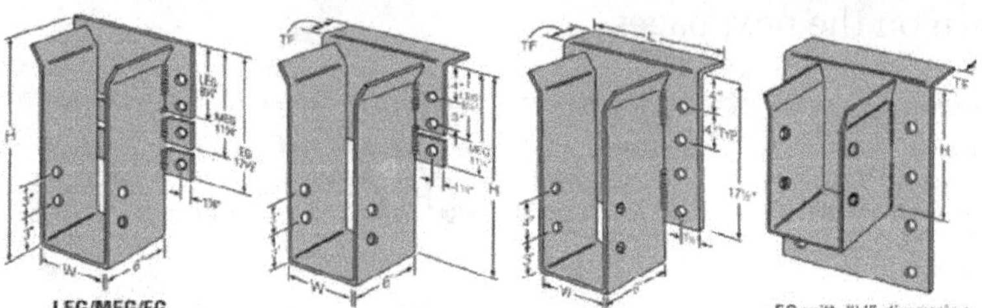

LEG/MEG/EG without Top Flange *(see options)*

LEG and MEG

EG

EG with "H" dimension less than the face plate height. The EG's back plate is always 17½", regardless of the stirrup height.

These products are available with additional corrosion protection.

Joist or Purlin Size	Model No.	Stirrup Ga	Dimensions W	Dimensions Min[3] H	Dimensions TF	Min. Header Depth	Bolts Header Qty	Bolts Header Dia	Bolts Joist Qty	Bolts Joist Dia	Without Top Flange Floor (100)	Without Top Flange Roof (125)	Top Flange No Triangle Theory Floor (100)	Top Flange No Triangle Theory Roof (125)	Top Flange Triangle Theory Floor (100)	Top Flange Triangle Theory Roof (125)	Code Ref.
3⅛ LAM	LEG3	7	3¼	9	2½	10	4	¾	2	¾	3465	4330	12675	13215	13040	13865	I19, F18, L14
5⅛ LAM	LEG5	7	5¼	9	2½	10	4	¾	2	¾	3465	4330	16290	16290	13040	13865	
	MEG5	7	5¼	9	2½	13	6	¾	2	¾	5170	6460	19710	19710	14835	16860	
	EG5	7	5¼	11	2½	20	8	1	2	1	8870	11085	20895	21815	17885	19865	
6¾ LAM	LEG7	7	6⅞	9	2½	10	4	¾	2	¾	3465	4330	16290	16290	13040	13865	
	MEG7	7	6⅞	9	2½	13	6	¾	2	¾	5170	6460	19710	19710	14885	16060	
	EG7	7	6⅞	11	2½	20	8	1	2	1	8870	11085	25320	25835	19290	21275	
8¾ LAM	EG9	7	8⅞	11	2½	20	8	1	2	1	8870	11085	25320	25835	20080	22875	

Model	Top Flange Ga	Top Flange Length (L)
LEG/MEG	7	12
EG5	3	11¾
EG7		13½
EG9		15½

Chapter 3

Tension Members

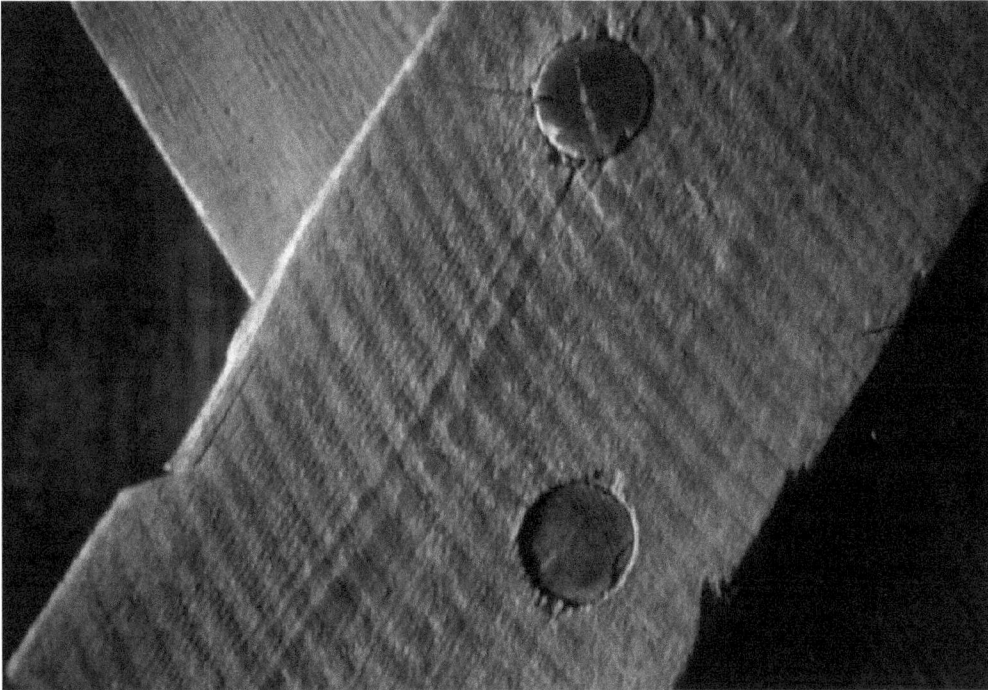

Axially loaded tension members can occur as a result of:

- Gravity-dominated loads such as hung weights.
- Chord members in roof diaphragms.
- Strut (collector) members in shear walls.
- Shear end Holdown members.
- Tension members in trusses.

The tension capacity of wood members is defined in below equation.

$$F_t' = F_t(C_D C_M C_t C_F)$$

Example 17:

Size the applicable vertical storage racks for the members shown below.

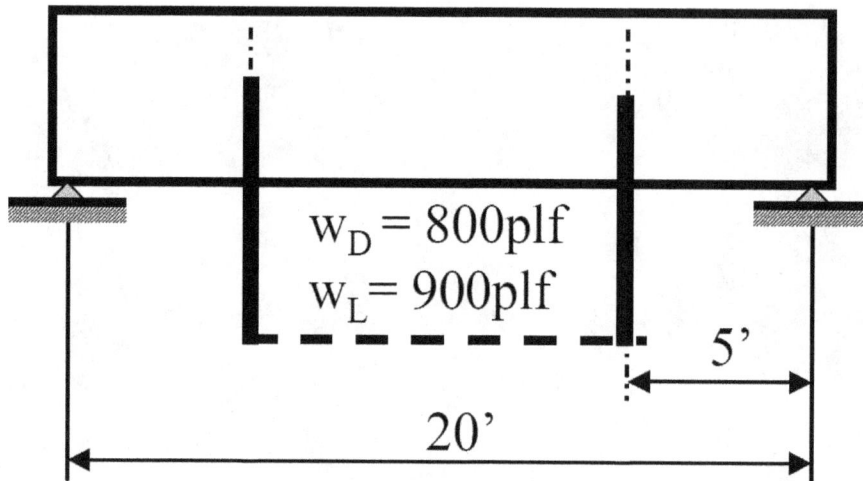

Dead Load: $P_D = 800 \times 10 \times 0.5 = 4{,}000$ lbs

Live Load: $P_L = 900 \times 10 \times 0.5 = 4{,}500$ lbs

Assume that connection to the beam will be made with a single row of 3/4"diameter

bolts.

Solution:

1. ASD Load Combinations: (use duration factor to estimate governing loading case):

    a) $D = \dfrac{4{,}000 \; lbs}{0.9} = 4{,}444 \; lbs$

    b) $D + L = \dfrac{(4{,}000 + 4{,}500) lbs}{1.0} = 8{,}500 \; lbs \therefore$ Governs Design

2. Estimate section dimensions:

Assume all other adjustment factors are 1.00 and $C_F = 1.30$):

$$F_{max} = 85{,}00 \; lbs$$

$$F_t = 675 \; psi$$

$$F_t' = 675 \ psi$$

$$A = \frac{8,500}{877.5} = 9.7 \ in^2$$

Add area of the holes $A_h = (0.75 + 0.0625)(1.5) = 1.22 \ in^2$

$$A_{required} = 9.7 + 1.22 = 10.92 \ in^2$$

Try 2x10 DF-L No.1 $(A_{2X10} = 13.88 \ in^2$

3.   Check Design

$C_F = 1.1$

$$F_t' = 675 \times 1.10 = 742.5 \ psi$$

The loss in the cross-section due to the oversized holes: $D' = D + \frac{1}{16}$

$$A_{available} = 13.88 - 1.22 = 12.66 \ in^2$$

$$f_t = \frac{8,500}{12.66} = 671 \ psi < 742.5 \ psi \ OK!$$

Therefore,

Use 2x10 DF-L No. 1

Example 18:

Size the applicable vertical storage racks for the members shown on next page.

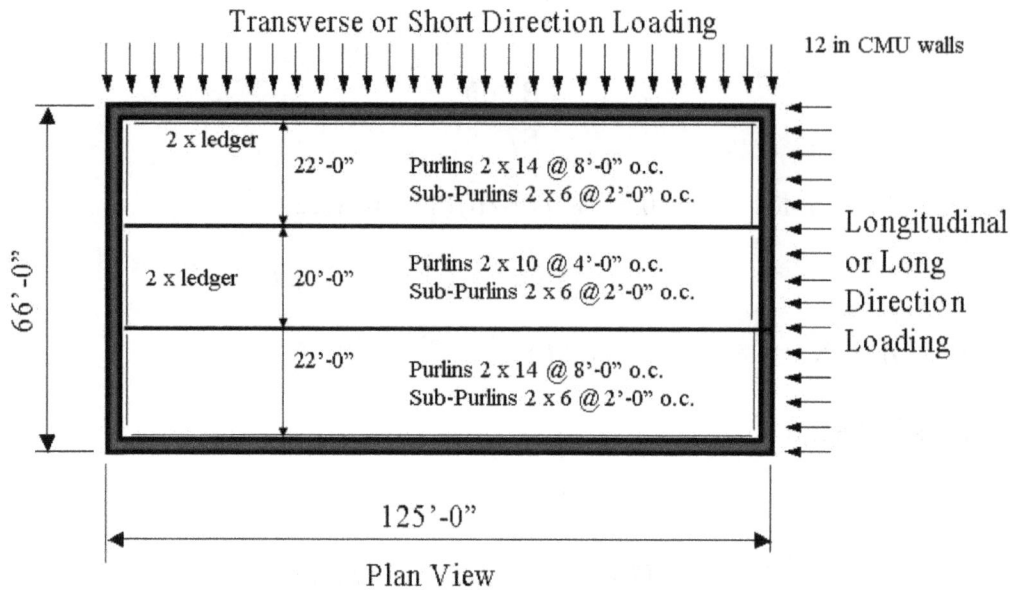

Transverse or Short Direction Loading

12 in CMU walls

2 x ledger
22'-0" Purlins 2 x 14 @ 8'-0" o.c.
Sub-Purlins 2 x 6 @ 2'-0" o.c.

2 x ledger 20'-0" Purlins 2 x 10 @ 4'-0" o.c.
Sub-Purlins 2 x 6 @ 2'-0" o.c.

22'-0" Purlins 2 x 14 @ 8'-0" o.c.
Sub-Purlins 2 x 6 @ 2'-0" o.c.

66'-0"

Longitudinal or Long Direction Loading

125'-0"

Plan View

$W_{Longitudinal\ Direction} = 800\ plf$ (Earthquke Controls)

$W_{Transverse\ Direction} = 400\ plf$ (Earthquke Controls)

Assume that connection of the wooden ledger to the CMU wall in the longitudinal wall will be made with two rows of ½″diameter bolts.

Assume that connection of the wooden ledger to the CMU wall in the transverse wall will be made with a single row of ½″diameter bolts.

Use exterior walls that are 16hx8bx12w CMU that is the width is 12 in.

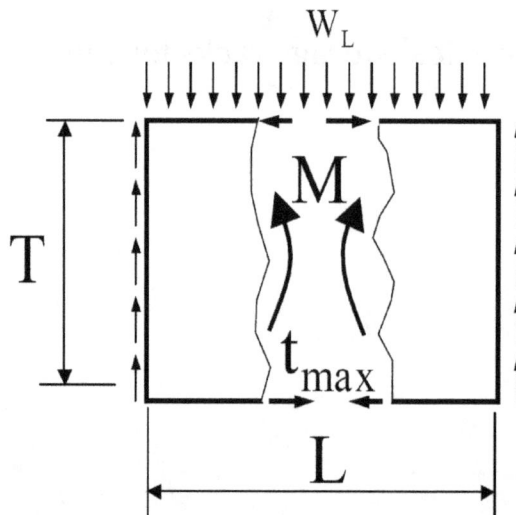

W_L

T

M

t_{max}

L

Solution:

1. Chord force in longitudinal direction:

   a) $M_L = \frac{W_L(L^2)}{8} = \frac{400(125^2)}{8} = 781.25$ kips-ft

   b) $F_{L,max} = \frac{M}{L_t} = \frac{781.25}{66} = 12\ kips$

2. Chord force in transverse direction:

   a) $M_T = \frac{W_T(L^2)}{8} = \frac{800(125^2)}{8} = 435.6$ kips-ft

   b) $F_{L,max} = \frac{M}{L_t} = \frac{435.6}{125} = 3.48\ kips$

3. The design of chord in longitudinal direction:

$C_D = 1.6$ and $C_F = 1.3$ all other factors are 1.0

$F_T = 675\ psi$

$F_t' = 675 \times 1.6 \times 1.3 = 1,404\ psi$

$A = \frac{12,000}{1,404} = 8.54\ in^2$

$A_{holes} = 2\left(0.5 + \frac{1}{16}\right)(1.5) = 1.7\ in^2$

$A_{required} = 8.54 + 1.7 = 10.24\ in^2$

Try 2x10 DF-L No. 1

$$A_{2x10} = 1.5 \times 9.25 = 13.88\ in^2\ (OK)$$

4. The design of chord in transverse direction:

$C_D = 1.6$ and $C_F = 1.3$ all other factors are 1.0

$F_T = 675\ psi$

$F_t' = 675 \times 1.6 \times 1.3 = 1,404\ psi$

$A = \frac{3,480}{1,404} = 2.48\ in^2$

$A_{holes} = 1\left(0.5 + \frac{1}{16}\right)(1.5) = 0.85\ in^2$

$A_{required} = 2.48 + 0.85 = 3.33\ in^2$

Try 2x4 DF-L No. 1 (Minimum allowable size)

$$A_{2x4} = 1.5 \times 3.5 = 5.25\ in^2\ (OK)$$

5. Check design of chord in longitudinal direction (Try 2x10)

$C_D = 1.6$ and $C_F = 1.1$ all other factors are 1.0

$F_T = 675 \; psi$

$F'_t = 675 \times 1.6 \times 1.1 = 1,188 \; psi$

$$A_{required} = 2(0.5 + \frac{1}{16})(1.5) = 1.70 \; in^2$$

$A_{available} = 13.88 - 1.70 = 12.18 \; in^2$

$f_t = \dfrac{12,000}{12.18} = 986 \; psi < 1,188 \; psi \; OK!$

USE 2x10 DF-L No. 1 for the Longitudinal Ledger

6. Check Design Chord Transverse Direction (Try 2x6)

$C_D = 1.6$ and $C_F = 1.3$ all other factors are 1.0

$F_T = 675 \; psi$

$F'_t = 675 \times 1.6 \times 1.3 = 1,404 \; psi$

$A_{holes} = 1\left(0.5 + \dfrac{1}{16}\right)(1.5) = 0.85 \; in^2$

$A_{available} = 5.25 - 0.85 = 4.40 \; in^2$

$f_t = \dfrac{3,480}{4.4} = 791 \; psi < 1,404 \; psi \; OK$

USE 2x4 DF-L No. 1 for the Transverse Ledger

COMBINED TENSION + BENDING MEMBERS

Adjusted tension stress capacity:

$$F'_t = F_t(C_D C_M C_t C_F)$$

Adjusted bending stress capacity:

$$F'_b = F_b(C_D C_M C_t C_L C_F C_{fu} C_i C_r)$$

Allowable combined stress capacity:

$$\frac{f'_t}{F'_t} + \frac{f'_b}{F'_b} \leq 1.0$$

Example 19:

Design a solid Sawn wood beam for the given conditions:

1- Use Select Structural. DF-L.

2- L=8ft

3- L_u = 0.00 ft. (beam is fully braced for out of plane buckling)

4- M_C < 19%.

5- W_D=250plf

6- W_{lr}=300plf

7- W_S=250plf

8- P_E=-8,000lbs (Tension from an earthquake strut force)

Solution:

ASD Load Combinations: (use duration factor to estimate governing loading case):

$$D = \frac{250}{0.9} = 278\ plf$$

$$D + L_r = \frac{250+300}{1.25} = 440\ plf\ \text{(Governs)}$$

$$D + S = \frac{250 + 250}{1.15} = 434\ plf$$

Estimate Section Dimension (use only the bending stress)

Note: assume all other adjustment factors are 1.00 to the exception of $C_D = 1.25$ and $C_F = 1.30$

$$M_{max} = \frac{WL^2}{8} = \frac{\left(\frac{550}{12}\right)(8 \times 12)^2}{8} = 52,800\ lbs - in.$$

$$F_b = 1,500\ psi\ \text{(Table 4A) DF-L Select Structural}$$

$$F_b' = F_b\ (C_D C_M C_t C_L C_F C_{fu} C_r) = 1,500\ x\ 1.25\ x\ 1.30\ x\ 1.00 = 2,473\ psi$$

$$S_{xx(req)} = \frac{52,800}{2,437} = 22\ in^3\ \text{(Required)}$$

Try a 4 x 8

$$S_{xx(4x8)} = 30.66 \ in^3$$

Check Design – Bending only

$$C_F = 1.3$$

$$F_b' = 2{,}473 \ psi$$

$$\rho = 62.4 \left(\frac{0.5}{1 + G(0.009)(19)}\right)\left(1 + \frac{19}{100}\right) = 34.20 \ \text{pcf/ft}$$

$$D = 6.168 \ plf$$

$$M_{max} = \frac{WL^2}{8} = \frac{(6.168)(8)^2}{8}(12) + 52{,}800 = 53{,}392 \ lbs - in.$$

$$f_b = \frac{M_{max}}{S_{xx}} = \frac{53{,}392}{30.66} = 1{,}585 \ psi < 2{,}437 \ psi \ OK!$$

Check Design – Bending with Combined Tension:

$F_t = 1{,}000 \ psi$

$F_t' = F_t(C_D C_M C_t C_F) = 1{,}000(1.6)(1.3) = 2{,}080 \ psi$

$F_b' = F_b \ (C_D C_M C_t C_L C_F C_{fu} C_r) = 1{,}500 \ x \ 1.6 \ x \ 1.30 \ x \ 1.00 = 3{,}120 \ psi$

$f_t' = \dfrac{8{,}000}{25.38} = 315 \ psi$

$f_b' = \dfrac{M}{S_x} = 1{,}585 \ psi$

$$\frac{f_t'}{F_t'} + \frac{f_b'}{F_b'} \leq 1.0$$

$$\frac{315}{2{,}080} + \frac{1{,}585}{3{,}120} = 0.66 \leq 1.0 \ \text{OK}$$

USE 4 x 8 DF-L Sel. Str.

Chapter 4

Compression Members

An orthotropic material has three mutually orthogonal twofold axes of rotational symmetry so that its material properties are, in general, different along each axis.

A familiar example of an orthotropic material is wood. In the wood design, one can define three mutually perpendicular directions. These are the axial direction (along the grain), the radial direction, and the circumferential direction, which is of extreme importance for axial compression of wood members. The new NDS 2015 is less conservative and leads to a more economical design.

Axially loaded compression members can occur as a result of:

- Gravity-dominated loads such as hung weights.
- Chord members in roof diaphragms.
- Strut (collector) members in shear walls.
- Shear end Hold down members.
- Compression members in trusses.

Adjusted Allowable Bending stresses in columns:

$$F_c' = F_c(C_D C_M C_F C_P C_t C_i)$$

C_D is the Load duration factor, which converts normal duration design values for other load duration values.

C_M is the wet service factor $\begin{cases} MC \leq 19\% \ Sawn \ Lumber \\ MC \leq 16\% \quad Glu \ Lams \end{cases}$

C_F is the size $\begin{cases} \quad\quad 1.0 \quad\quad\quad\quad MSR \ and \ MEL \\ \quad\quad 1.0 \quad\quad\quad\quad\quad GLBs \\ See \ the \ table \ 4 \ for \ solid \ sawn \end{cases}$

C_P is the column stability factor, and is equal to 1 for a fully supported column:

$$C_p = \frac{1 + \frac{F_{CE}}{F_C^*}}{2c} - \sqrt{\left(\frac{1 + \frac{F_{CE}}{F_C^*}}{2c}\right)^2 - \frac{\frac{F_{CE}}{F_C^*}}{c}}$$

$\frac{F_{CE}}{F_C^*} = Q$ (F_{CE} is the smaller of F_{CEx} and F_{CEy}.

$$C_p = \frac{1 + Q}{2c} - \sqrt{\left(\frac{1 + Q}{2c}\right)^2 - \frac{Q}{c}} \leq 1.0$$

$$c = \begin{cases} 0.8 & \text{For sawn lumber} \\ 0.85 & \text{Visually graded lumber} \\ 0.9 & \text{Glu lam members} \end{cases}$$

$F_{CEx} = \frac{0.822 E'_x}{\left(\frac{l_{ex}}{b}\right)^2}$ Euler critical stress for columns buckling about the X-axis.

$F_{CEy} = \frac{K_{cE}E'_y}{\left(\frac{l_{ey}}{b}\right)^2}$ Euler critical stress for columns buckling about the X-axis

$E'_y = E_y(C_M C_t)$ Properties about the y-axis (Strong Axis)

$E'_x = E_x(C_M C_t)$ Properties about the x-axis (Weak Axis)

The NDS 2015 specifications allows the use of $E'_x = E'_y = E'_{min}$

$$K_{cE} = 0.822$$

$$l_e = K_e l_u < 50$$

K_e: Column effective buckling length factor

l_{uy} length of the unbraced column for buckling about the Y-axis

l_{ux} length o the unbraced column for buckling about the X-axis

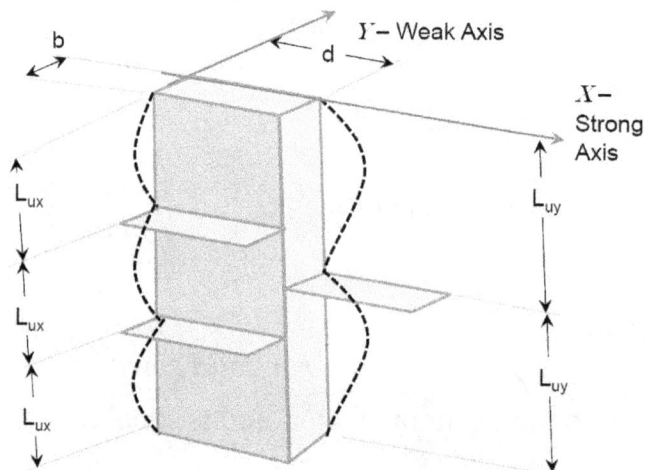

Table 17 Buckling length coefficients (NDS 2015)

Buckling modes						
Theoretical K_e value	0.5	0.7	1.0	1.0	2.0	2.0
Recommended design K_e when ideal conditions approximated	0.65	0.80	1.2	1.0	2.10	2.4
End condition code		Rotation fixed, translation fixed				
		Rotation free, translation fixed				
		Rotation fixed, translation free				
		Rotation free, translation free				

Example 20:

Size the applicable square member for the following.

- L=10ft (simply supported both ends).

- MC <19%

- Member not treated or incised. Member is Visually Graded (VGL)

- Load is $D + L_r, P = 15 \ Kips$

- This is a simply supported square column.

- Bracing is provided as in the below figure

- Use DF-Larch Select Structural and select a square member

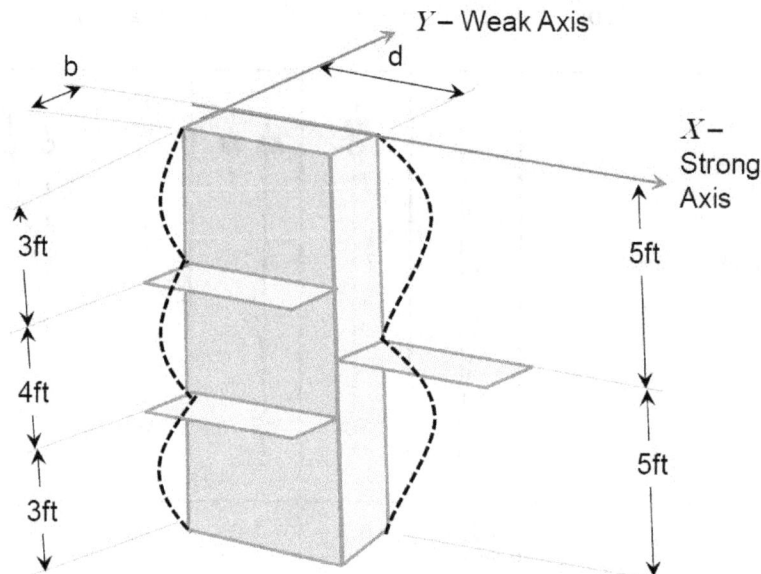

$E_y = 1,600,000\ psi$ (Strong axis bending)

$E_x = 580,000\ psi$ (Weak axis bending)

Since this is a square member cannot ensure member will be installed in one particular direction. Use $E_{min} = 580,000\ psi$

$b = d$ (Must use a square member as indicated in the problem statement)

Note: Table 4D of NDS (NDS 2015) shows the design values for visually graded timbers.

Solution:

Step 1: Select first the trial size (Assume a size greater than 5x5)

$F_c = 1,150\ psi$ (compression parallel to grain to be used in posts)

$E_x = 280,000\ psi$ (Note: This is the minimum E and member is square use this one.)

$E_y = 1,600,000\ psi$

$$C_M = C_t = 1$$

$$C_D = 1.25$$

$$F_c^* = F_c C_D = 1,150(1.25) = 1,437.5 \, psi$$

Assume $C_p = 0.25$ as a first trial (This is just a preliminary estimate) Student is encouraged to keep a table of C_p values for use in the future, for trial sizes estimates.

Trial Area use $A = \frac{P}{F_c^* C_p}$

$$A = \frac{15,000}{(1,437.5)0.25} = 41.8 \, in^2$$

Could select a 7x7 but this section does not exist

So try a square section use an 8x8 member

$$A_{8x8} = 7.5 \times 7.5 = 56.25 \, in^2$$

Step 2 Check trial size: A_{8x8} (Size greater than 5x5 use the design tables)

$F_c = 1,150 \, psi$ (**Note: Compression parallel to grain and member greater than 5"x5"**)

$$C_{Fx} = \left(\frac{12}{7.5}\right)^{\frac{1}{9}} = 1.05$$

$$C_{Fy} = \left(\frac{12}{7.5}\right)^{\frac{1}{9}} = 1.05$$

$$F_c^* = 1,150(1.25)(1.05) = 1,509 \, psi$$

$$K_e = 1.0 \, (Simply \, supported \, beams)$$

$$L_{ux} = K_e(4)(12) = 48 \, in$$

$$L_{uy} = K_e(5)(12) = 60 \, in$$

$$\frac{L_{ux}}{b} = \frac{48}{7.5} = 6.4 < 50$$

$$\frac{L_{uy}}{d} = \frac{60}{7.5} = 8 < 50$$

$$F_{cEx} = \frac{(0.822)(580,000)}{6.4^2} = 11,639 \; psi$$

$$F_{cEy} = \frac{(0.822)(580,000)}{0.8^2} = 7,449 \; psi$$

$$Q_x = \frac{11,639}{1,094} = 7.71$$

$$Q_y = \frac{7,449}{1,509} = 4.93$$

$$c = 0.85 \text{ Visually graded wood}$$

$$C_p = \frac{1 + 4.93}{2(0.85)} - \sqrt{\left(\frac{1 + 4.93}{2(0.85)}\right)^2 - \frac{4.93}{0.85}} = 0.96 \le 1.0$$

$$F_c^* = 1,509(0.96) = 1,456 \; psi$$

$$A_r = \frac{15,000}{1,456} = 10.30 \; in^2$$

Square member try a new size 4x4 $A = 12.25 \; in^2$

Step 3: Check another trial size: A_{4x4}

$F_c = 1,700 \; psi$ (**Note: Compression parallel to grain and but member NOT greater than 5"x5" material properties change**)

$E_x = 69,000 \; psi$ (weak axis bending)

$$\text{Use } E_{min} = 69,000 \; psi$$

$$C_{Fx} = 1.15$$

$$C_{Fy} = 1.15$$

$$F_c^* = F_c C_D = 1,700(1.25)(1.25) = 2,443 \ psi$$

$$K_e = 1.0 \ \text{(Simply supported member)}$$

$$L_{ux} = K_e(4)(12) = 48 \ in.$$

$$L_{uy} = K_e(5)(12) = 60 \ in.$$

$$\frac{L_{ux}}{b} = \frac{48}{3.5} = 13.7 < 50 \ ok$$

$$\frac{L_{uy}}{d} = \frac{60}{3.5} = 17.1 < 50 \ ok$$

$$F_{cEx} = \frac{(0.822)(690,000)}{13.7^2} = 3,015 \ psi$$

$$F_{cEy} = \frac{(0.822)(690,000)}{17.1^2} = 1,929 \ psi$$

$$Q_x = \frac{3,015}{2,443} = 1.23$$

$$Q_y = \frac{1,929}{2,443} = 0.78$$

$$c = 0.85 \ \text{Visually graded wood}$$

$$C_p = \frac{1 + 0.78}{2(0.85)} - \sqrt{\left(\frac{1 + 0.78}{2(0.85)}\right)^2 - \frac{0.78}{0.85}} = 0.62 \le 1.0$$

$$F_c^* = 2,443(0.62) = 1,538 \ psi$$

$$A_r = \frac{15,000}{1,538} = 9.75 \ in^2$$

Arzhang Zamani

Square member size 4x4 $A = 12.25\ in^2$ OK

USE braced 4x4 DF-L Select Structural

4.1 Combined Bending and Compression Loads

- Axially loaded compression members can occur because of:
  a) Columns Supporting Beams
  b) Chord members in roof diaphragms.
  c) Strut (collector) members in shear walls.
  d) Shear end Hold down members.
  e) Compression members in trusses.

Adjusted Compression capacity:

$$F_c' = F_c\ (C_D C_M C_t C_F C_p)$$

Adjusted bending stress capacity:

$$F_b' = F_b\ (C_D C_M C_t C_L C_F)$$

Allowable combined stress capacity:

$$\left(\frac{f_c}{F_c'}\right)^2 + \frac{f_{b1}}{F_{b1}'\left(1 - \frac{f_c}{F_{cE1}}\right)} \le 1.0$$

$w_{D+S} = 56.20 \, pli$

L = 13'-0"

$w_{D+S} = 56.20 \, pli$

$P_C = 28,247 \; lbs$ $P_C = 28,247 \, lbs$

L = 13'-0"

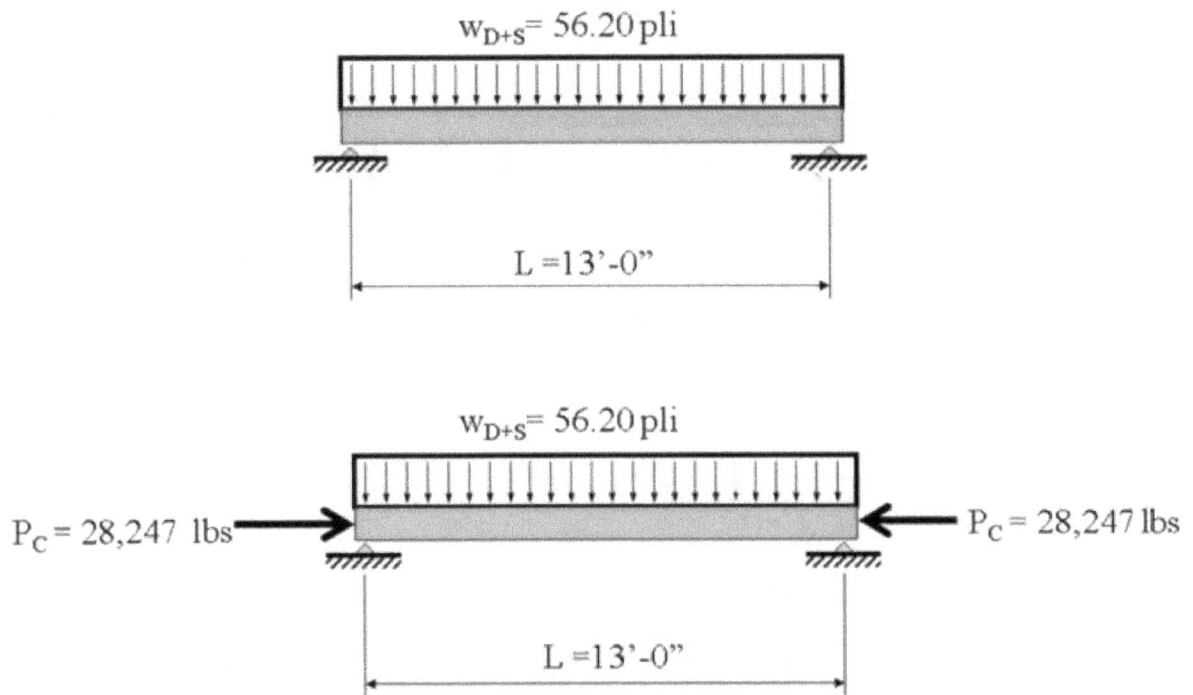

Chord and strut forces:

A. Forces that develop from lateral loads in the transverse or short direction of loading are given.

B. Forces that develop from lateral loads in the longitudinal or long direction are similar, but the orientation is reversed.

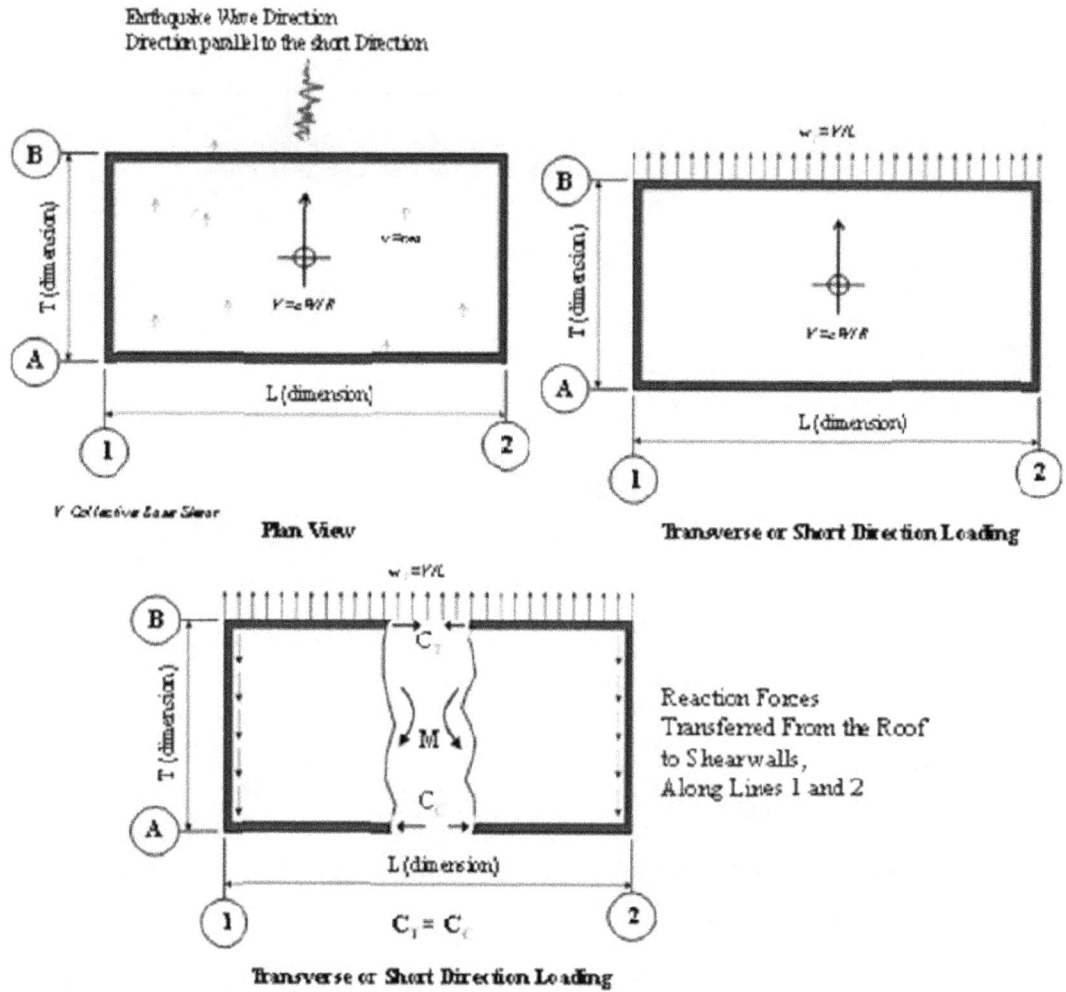

Plan View

Transverse or Short Direction Loading

Transverse or Short Direction Loading

Chord forces, C_T and C_c, along lines A and B originated from lateral loads W_T and are given by:

$$M = \frac{W_t L^2}{8}$$

$$C_T = C_C = \frac{M}{T}$$

Strut forces along lines 1 and 2 are given by:

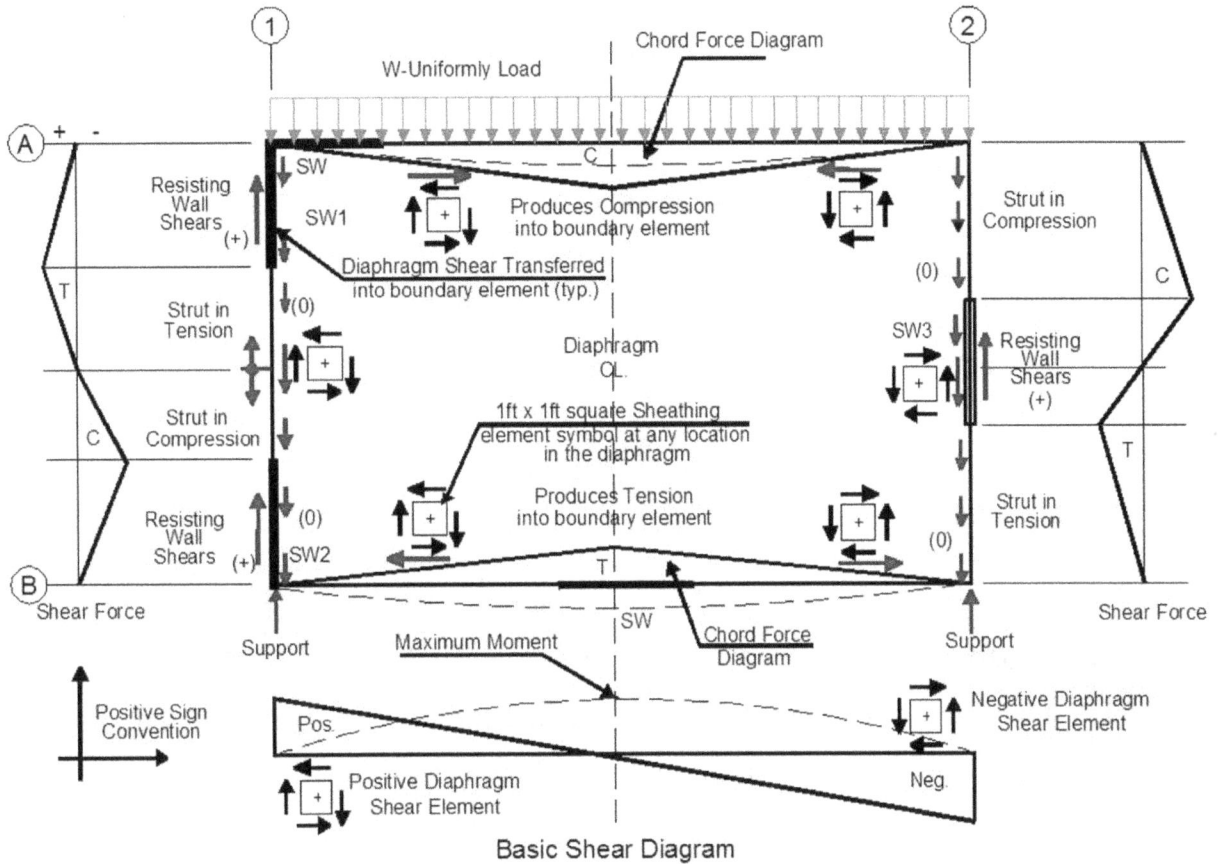

Shear Distribution Into a Simple Diaphragm

- Calculate the reaction forces along lines 1 and 2:

$$R_T = \frac{W_t L}{2}$$

- Calculate the total length of roof length along lines 1 and 2

$$T_R = \sum_{i=1}^{n} T_{R,i}$$

- Compute the shear load per unit length of roof along line 1 and 2

$$S_R = R_T \frac{1}{T_R}$$

- Calculate the total length of shear wall in each direction 1 and 2

$$T_{SW} = \sum_{i=1}^{n} T_{SW,i}$$

- Compute the reacting shear load per unit length of wall along line 1 and 2

$$S_{SW} = R_T \frac{1}{T_{SW}}$$

Calculate the strut forces per:

$$S_{T,i} = (S_R - S_{SW})(x_i - x_{i-1}) + S_{T,x-i}$$

Example 21:

Using the information provided in the figures below design one of the headers shown in the floor plan. Use moisture content for 2xs or smaller of 18%, Use moisture content for 4xs or larger of 16%, Use a moisture content for GLB of 15% and a weight of 36 pcf.

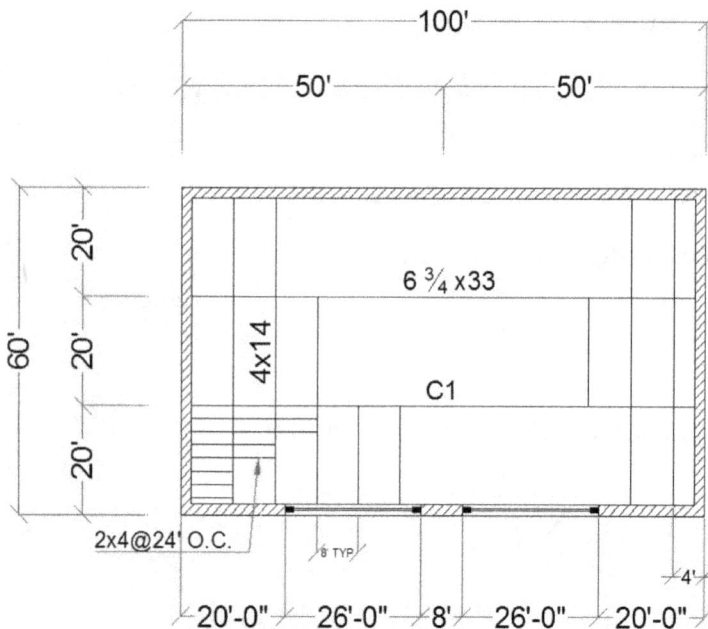

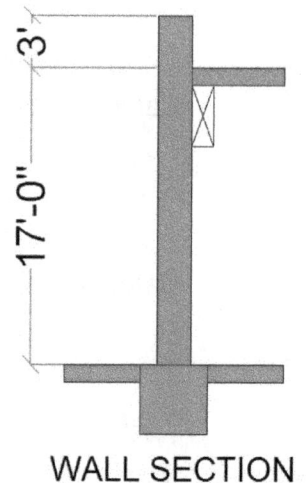

WALL SECTION

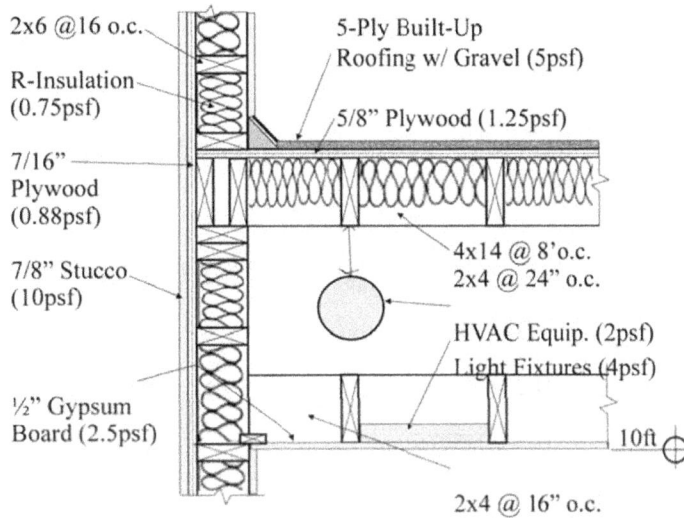

2x6 @16 o.c.

R-Insulation (0.75psf)

7/16" Plywood (0.88psf)

7/8" Stucco (10psf)

½" Gypsum Board (2.5psf)

5-Ply Built-Up Roofing w/ Gravel (5psf)

5/8" Plywood (1.25psf)

4x14 @ 8'o.c.
2x4 @ 24" o.c.

HVAC Equip. (2psf)
Light Fixtures (4psf)

10ft

2x4 @ 16" o.c.

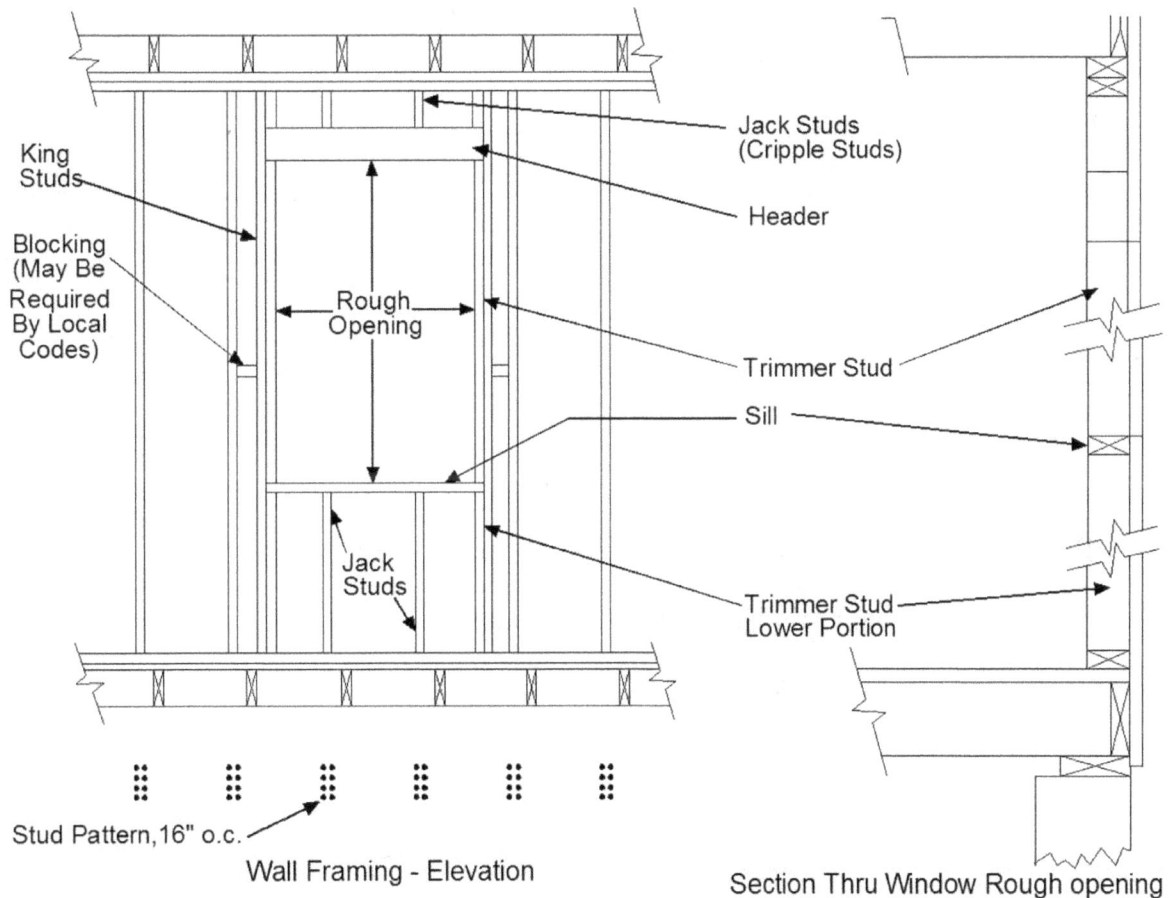

King Studs

Blocking (May Be Required By Local Codes)

Rough Opening

Jack Studs

Stud Pattern, 16" o.c.

Jack Studs (Cripple Studs)

Header

Trimmer Stud

Sill

Trimmer Stud Lower Portion

Wall Framing - Elevation

Section Thru Window Rough opening

Note: Position base of the header at 11 ft.

The design of the header beam must use the following load values:

1. The Roof live load is 20psf.

2. The Snow Load is 35psf.

3. Wind forces are for a qs = 16.0 psf (based on a 100mph).

$C_e \Rightarrow$	$Exposure\ B \rightarrow \begin{cases} 0.62 & (0' \le h \le 15') \\ 0.67 & (15' \le h \ge 20') \end{cases}$ $(UBC\ Table16-G)$
$C_q \Rightarrow$	$\begin{cases} 0.80 & windward \\ -0.50 & leeward \\ -0.70 & roof \end{cases}$ Assuming » flat roof and closed building
$I_W \Rightarrow$	1.00

4. The seismic forces will be applied in terms of the following base shear considerations: V=1.25W/R and R=4.

DESIGN SOLUTION:

Strategy:

1. Roof base dead load

2. Framing members dead load

3. Rood dead load and separate it in weights with and w/out the GLBs.

4. Wall dead load

5. Building Total Load (including the GLBs weight)

6. Seismic Base Shear and Uniform Design loads

7. Wind Uniform Design loads

8. Chord Forces

9. Strut Forces

10. Design the header for combined loads

Design Loads for the Header Beam:

Note: Use a wall frame with the plywood roof sheathing going all the way to the exterior edge of the walls. This leads to a tributary length for the header design from the roof loads of $L_t = \frac{20}{2} = 10 \ ft$.

Summary of Dead Loads for the header:

1. Weight of the tributary roof loads (These are computed in the next page)

a. Dead with GLBs = 20.27 psf (for use only in the seismic analysis, **V**)

b. Dead without GLBs = 18.41 psf (for use in design of the header)

2. Weight of the wall above the header

a. Walls Base Dead = 15.59 psf

3. Self- Weight of the Header to be incorporated later in the design.

Dead Load		
Built Up Roofing	5.00	
5/8 Plywood	1.25	
Insulation	0.75	
HVAC	2.00	
Misc + Lights	4.00	
½" Gypsum Board	2.50	
Roof Base	**15.50**	psf
Roof Purlins		
4x14 spacing	8.00	ft
b	3.50	in
d	13.25	in
G	0.50	
mc	16.000	%
rho	33.76	pcf
A	46.38	in2
DL 4x14	**1.36**	psf
Roof subPurlins		
2x4 spacing	24.00	in
b	1.50	in
d	3.50	in
G	0.50	
mc	18.000	%
rho	34.06	pcf
A	5.25	in2
DL 2x4 subpurlins	**0.62**	psf
Ceiling Framing		
2x4 spacing	16.00	in
b	1.50	in
d	3.50	in
G	0.50	
mc	18.000	%
rho	34.06	pcf
A	5.25	in2
DL 2x4 ceiling	**0.93**	psf

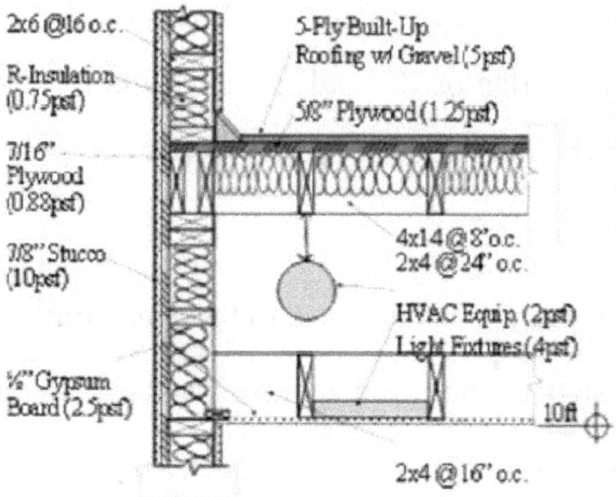

Diagram labels:
2x6 @16 o.c.
R-Insulation (0.75psf)
7/16" Plywood (0.28psf)
7/8" Stucco (10psf)
½" Gypsum Board (2.5psf)
5-Ply Built-Up Roofing w/ Gravel (5psf)
5/8" Plywood (1.25psf)
4x14 @ 8' o.c.
2x4 @ 24' o.c.
HVAC Equip (2psf)
Light Fixtures (4psf)
2x4 @ 16' o.c.
10ft

Dead Load		
Roof Base	15.50	psf
4x14 Purlins	1.36	psf
2x4 subpurlins	0.62	psf
2x4 Celing Joists	0.93	psf
Total Dead (no GLBs)	**18.41**	psf
GLB Dead Load		
b	6.75	in
d	33.00	in
A	222.75	in2
rho	36.00	pcf
Wglb	55.69	plf
Ltotal	200.00	ft
Ptotal	11,138	lbs
Roof L	100	ft
Roof W	60	ft
Roof A	6,000	ft2
W GLBs	**1.86**	psf
Total Roof Dead		
Total Dead (no GLBs)	18.41	psf
W GLBs	1.86	psf
Total Roof Dead	**20.27**	psf

Wall Dead Loads:

Dead Load		
7/8 Stucco	10.00	
7/16 Plywood	0.88	
Insulation	0.75	
½" Gypsum Board	2.50	
Wall Base	**14.13**	psf

Wall Studs		
2x6 spacing	16.00	in
b	1.50	in
d	5.50	in
G	0.50	
mc	18.000	%
rho	34.06	pcf
A	8.25	in2
DL 2x6 Studs	**1.46**	psf

Diagram labels:
- 2x6 @16 o.c.
- 5-Ply Built-Up Roofing w/ Gravel (5psf)
- R-Insulation (0.75psf)
- 5/8" Plywood (1.25psf)
- 7/16" Plywood (0.88psf)
- 7/8" Stucco (10psf)
- 4x14 @ 8' o.c.
- 2x4 @ 24" o.c.
- HVAC Equip. (2psf)
- Light Fixtures (4psf)
- ½" Gypsum Board (2.5psf)
- 10ft
- 2x4 @16" o.c.

Wall Load		
Wall Base	14.13	psf
2x6 studs Purlins	1.46	psf
Total Dead (no GLBs)	**15.59**	psf

Summary of Header Vertical Uniform Design Loads:

Dead Loads						
Roof Dead w/ GLBs	20.27	psf	Uniform Dead Load Roof	184.11	plf	
Roof Dead w/out GLBs	18.41	psf		15.34	pli	
Wall Dead	15.59	psf	Uniform Dead Load Walls	140.34	plf	
				11.70	pli	
			Total Dead Load	27.04	pli	
Header Design Loads			Uniform Live Load	200.00	plf	
LT	10.00	ft		16.67	pli	
H Top Parapet	20.00	ft	Uniform Snow Loads	350.00	plf	
H bot Header	11.00	ft		29.17	pli	
HT	9.00	ft				
Dead Roof	18.41	psf		CD		
Wall Dead	15.59	psf	D	27.04	0.9	30.04 pli
Live Load	20.00	psf	D+L	43.70	1.25	34.96
Snow Load	35.00	psf	D+S	56.20	1.15	48.87

Note: Self- Weight of the Header not yet included and is to be incorporated later in the design.

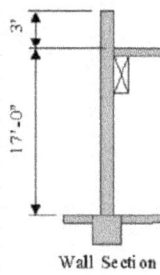

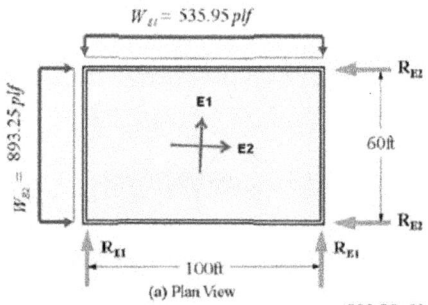

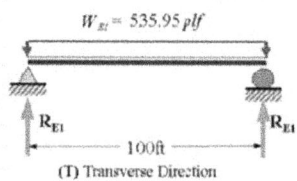

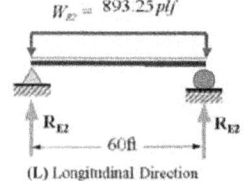

WIND CALCULATIONS

Wind Design Loads

Roof L	100.00	ft
Roof W	60.00	ft
Roof T	20.00	ft
Roof H	17.00	ft
Parapet H	**3.00**	ft

qs	16.00	psf	
Ce	0.62	15.00	
	0.67	20.00	
Cq	0.80	windward	
	-0.50	leeward	
I	1.00		

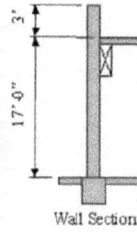

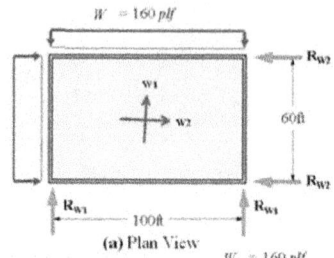

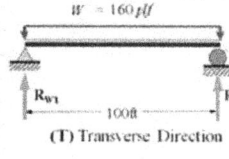

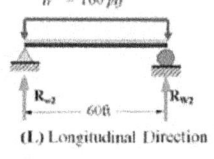

Calculations

$$P = C_e\, C_q\, q_s\, I_w$$

	windw (psf)	leew (psf)	P (psf)	Heigth (ft)
Load H2	8.60	-5.40	**14.00**	20.00
Load H1	7.90	-5.40	**13.30**	15.00
Wr	**160.07**	plf	load on the roof	

Calculations

	W1 (‖60)	W2 (100‖)	
L	100	60	ft
v	160.07	160.07	plf
Chord Moment	**506.20**	**445.51**	kips-ft
Chord Distance, H	60	100	
Chord Forces (T/C)	**8,437**	**4,455**	lbs

$$M = \frac{vL^2}{8}$$

$$C = M/H$$

Wall Section

(a) Plan View

(T) Transverse Direction   (L) Longitudinal Direction

$W = 160\,plf$

8.6+5.4 = 14.00 (lb/ft)/ft of wall

13.30 (lb/ft)/ft of wall

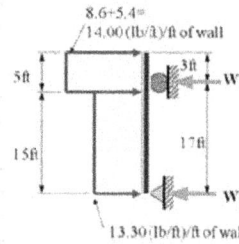

SEISMIC CALCULATIONS

Seismic Design Loads

Roof L	100.00	ft
Roof W	60.00	ft
Roof A	6,000	ft2
Roof Dead	20.27	psf
Total Roof Weigth	**121.60**	kips
Total Wall Heigth /2	10.00	ft
Total Wall Length (perimeter)	320.00	ft
Wall Dead	15.59	
Total Wall Weigth	**49.90**	kips
Total Building Weigth	**171.50**	kips

Base Shear

Ce	1.25	
W	171.50	kips
R	4.00	
V	**53.59**	kips

$$V = 1.25\frac{W}{R}$$

Calculations

	E1 (‖60)	E2 (100‖)	
L	100	60	ft
v	535.55	893.25	plf
Chord Moment	**1694.32**	**2486.05**	kips-ft
Chord Distance, H	60	100	
Chord Forces (T/C)	**28,247**	**24,860**	lbs

$$M = \frac{vL^2}{8}$$

$$C = M/H$$

Wall Section

(a) Plan View

(T) Transverse Direction   (L) Longitudinal Direction

$W_{E1} = 535.95\,plf$

$W_{E2} = 893.25\,plf$

Strut Forces Design Calculations:

SW1	20.00	ft
Op1	26.00	ft
SW2	8.00	ft
Op2	26.00	ft
SW3	20.00	ft
Total L	100.00	

Reactions Forces Along the wall		
W seismic	893.25	plf
Wind	160.07	plf
L	60.00	
Reaction S	**26.80**	kips	Governs
Reaction W	**4.80**	kips

Sheating Shear		
R govern	**26.80**	kips
L	100.00	
Sroof	**267.97**	plf

Total Shear Wall Length	48	
Swalls	**558.28**	plf

Force along the Wall	Sroof	Swalls	F Roof	F wall	Force
0	267.97	558.28	0	0	0
20	267.97	558.28	5,359	11,166	5,806
33	267.97	0.00	8,843	11,166	2,322
46	267.97	0.00	12,327	11,166	-1,161
54	267.97	558.28	14,471	15,632	1,161
67	267.97	0.00	17,954	15,632	-2,322
80	267.97	0.00	21,438	15,632	-5,806
100	267.97	558.28	26,797	26,797	0

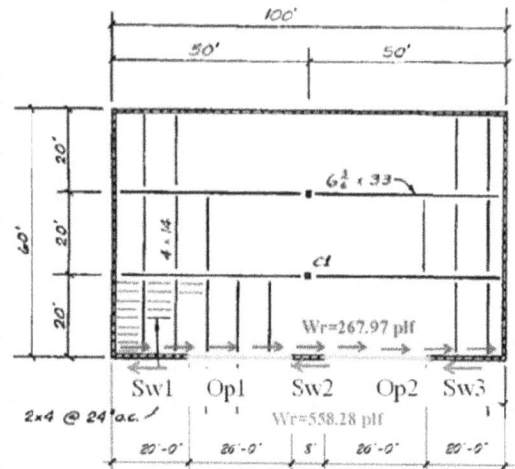

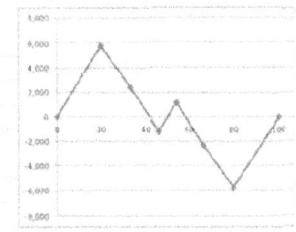

Header Load Summary

Dead Load

Possible Load Combinations	CD value	Load Values	Design Values
(1) D	0.90	W=27.04 pli	W=30.01 pli
(2) D + LR	1.25	W=43.70 pli	W=34.96 pli
(3) D + S	1.15	W=56.20 pli	W=48.87 pli
(4) D + W + LR	1.60	W=43.70 pli P= 8,437 lbs	W=27.32 pli P= 5,272 lbs
(5) D + W + S	1.60	W=56.20 pli P= 8,437 lbs	W=35.12 pli P= 5,272 lbs
(7) D + 0.7E + LR	1.60	W=43.70 pli P =28,247 lbs	W=27.32 pli P *=12,358 lbs
(8) D + 0.7E +S	1.60	W=56.20 pli P = 28,247 lbs	W=35.12 pli P *= 12,358 lbs

* Note P= 28,247*0.70/1.6=12,358 lbs

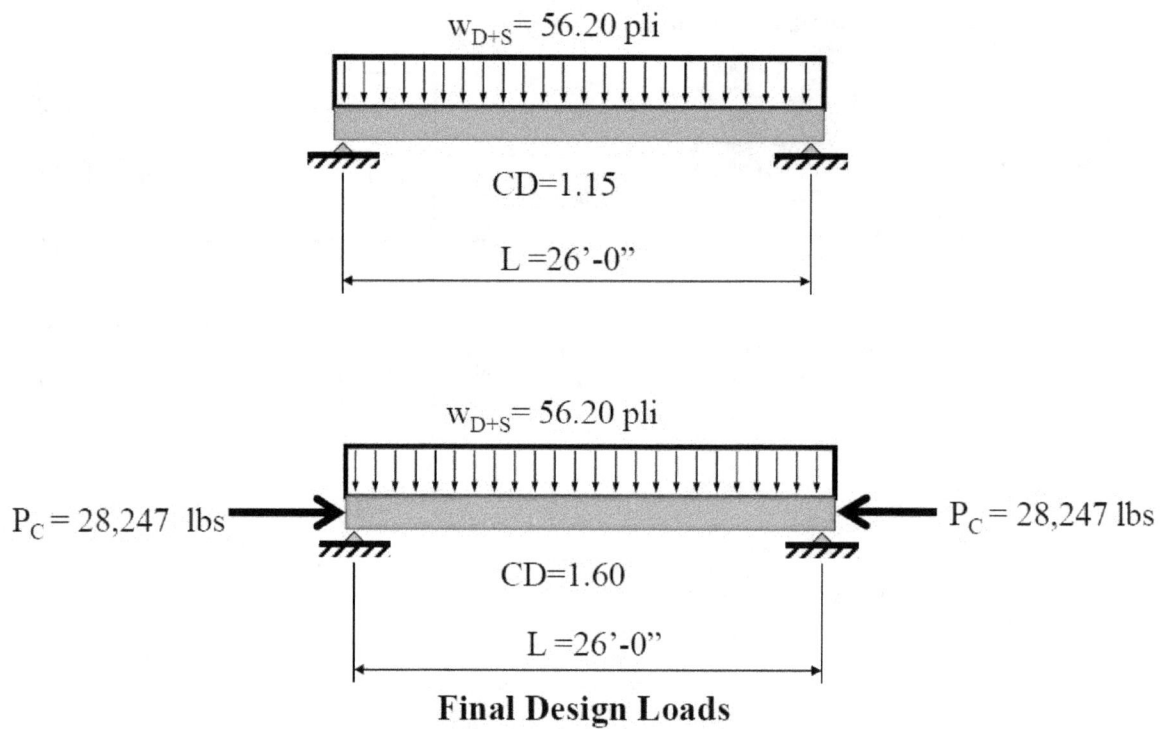

Final Design Loads

GIVEN PROBLEM STATEMENT

Wood Species	DF-L			
Beam Type	Select Stru			
MC limit	19	Note: 16 for GLB 19 Solid Sawn		
Moisture	15			

Deflection Limits

Ceiling type	Plaster	P
Deflection L	360	
Deflection D+L	240	

LOADING

			weigth	
wD	28.25	324.45	14.51	plf
wL	16.67	200	plf	
wS	29.17	350	plf	

1.21

Duration Factor

	ASD (D)	ASD (D+L)	ASD (D+S)	
CD	0.9	1.25	1.15	
Govern ASD ?	376.62	431.17	599.09	plf
			Governs ?	

Beam Dimensions

b	5.500	in	Try 6x12	
d	11.250	in	33.76	pcf

Beam Length

L	156.00	13.00	ft

Material Properties

Fb	1,500	psi
Fv	180	psi
Fc+	625	psi
E	1,900,000	psi

Adjustment Factors

	Fb	Fv	Fc+	E
CM	1.00	1.00	1.00	1.00
Ct	1.00			
CL	1.00	Need to check for supports later		
CF (not for GLB)	1.00			
xyz	20			
k	1			
CV (only in GLB)	1.00	Note: Do not use CL and CV together		
Cfu	1.00			
Cr	1.00			
Cc	1.00			
Cf	1.00			

Design For Bending

Mmax	174650	lbs-in
CD	1.15	
CM	1.00	
Ct	1.00	
CL	1.00	
CF	1.00	
CV	1.00	Note: Do not use CL and CV together
Cfu	1.00	
Cr	1.00	
Fb	1,500	psi
F'b	1,725	psi
b	5.500	in
d	11.25	in
Sx	116.02	in
fb	1,505	psi
check	OK	

Design For Shear

Vmax	4478	lbs
w	57.41	lbs/in
b	5.500	in
d	11.25	in
Area	61.88	in^2
V'	3832	lbs
CD	1.15	
CM	1.00	
Ct	1.00	
CH	1.00	
Fv	180	
F'v	207	psi
fv	109	psi
check	OK	

Deflection Limits

	Calculated	Limits Deflection		
L	0.18	0.43	in	OK
D+L	0.28	0.65	in	OK

Design For Bearing

R	4478	lbs
CM	1	
Ct	1	
Cb	1	Need to check supports later
Fc+	625	psi
F'c+	625	psi
b	5.500	in
lb	1.50	in
R cap	15785	lbs
Check Flat	OK	
Use a lb	2	in
fv	407	psi
Cb	1.19	
F'c+	742	psi
check	OK	

Design for bracing

lu-top	96.00	8	ft	Assume purlines/joists at 4ft on center.
lu-bottom	96.00	8	ft	max of l1 and l2 times 2
b	5.500	in		
d	11.25	in		

Design for stability

Ey	1,900,000	psi
	GLB	
Kbe	0.610	
lu/d top	8.53	
lu/d bottom	8.53	

$$l_e = \begin{cases} 2.06\, l_u & l_u/d < 7 \\ 1.63\, l_u + 3d & 7 \le l_u/d \le 14.3 \\ 1.84\, l_u & l_u/d > 14.3 \end{cases}$$

le top	190.23	
le bottom	190.23	
Rb top	8.41	OK
Rb bottom	8.41	OK

$$R_B = \sqrt{\frac{l_e d}{b^2}} \le 50$$

FbE-top	16382	psi
FbE-bottom	16382	psi
Fb*	1725	psi	Remove the CV since CL is not to be used with CL

$$F_{bE} = \frac{K_{bE} E_y'}{R_B^2}$$

Q-top	9.50	
Q-bottom	9.50	

$$Q = \frac{F_{bE}}{F b^*}$$

CL top	0.99	
CL bot	0.99	
Fb top	1715	psi
Fb bot	1715	psi

$$C_L = \left[\left(\frac{1+Q}{1.9}\right) - \sqrt{\left(\frac{1+Q}{1.9}\right)^2 - \frac{Q}{0.95}}\right] \le 1.0$$

fb	1505	psi
	OK	
	OK	

Try a 6x12 DF-L Select Structural for the combined loading design

Check combined loading design of a 6x12

						Comb. 3	
L	13.0	ft					
	Comb. 3		Comb. 8			Bending	
w 6x12	1.210		1.210	pli	Fb	1,500	
W D+S	57.41		57.41	pli	CD	1.150	
P C	0.00		28247.00	lbs	CF	1.000	
PC * 0.7	0.00		19772.90	lbs	Cr	1.000	
KcE	0.822				CL	1.000	
Ex	510,000	psi			Fb'	1725.00	
Ey	510,000	psi			w D+S	57.41	pli
b	5.500	in			M	174,655	lbs-in
d	11.250		Try a 6x12		Sx	116.02	in 3
					fb	1505.44	psi
					F'b	1725.00	psi
					fb	1505.44	psi
						OK	

$w_{D+S}= 56.20$ pli

CD=1.15

L =26'-0"

		Comb. 8				Bending	
		Compression			Fb	1,500	
	Fc	850			CD	1.600	
	CD	1.600			CF	1.000	
	CF	1.000			Cr	1.000	
	Cr	1.000			CL	1.000	
	Fc*	1360.00	psi		Fb'	2400.00	
		Col. Buckling Cp			fb	1505.44	psi
Ke		1.000				**OK**	
luy		96.000	8				
le/d		8.53	ratio ok			**Combined**	
	FcE	5757.11	psi		fc	456.52	psi
	Fc	1360.00	psi		F'c	1302.84	psi
	Qy	4.23			fb1	1505.44	psi
					F'b1	2400.00	psi
	c	0.850			FCE	5757.11	psi
	Q	4.23				0.80	
	Cp	0.96				**OK**	
	F'c	1302.84	psi				
	fc	456.52	psi				
		OK					

Use a 6x12 DF-L Select Structural Header Beam

For Completeness check tension plus bending

L	13.0	ft
	Comb. 3	Comb. 8	
w 6x12	1.210	1.210	pli
W D+S	57.41	57.41	pli
P T	0.00	28247.00	lbs
PT * 0.7	0.00	19772.90	lbs
KcE	0.822		
b	5.500	in
d	11.250	Try a 6x12

$w_{D+S} = 56.20$ pli

CD=1.15

L =26'-0"

$w_{D+S} = 56.20$ pli

$P_T = 28,247$ lbs $P_T = 28,247$ lbs

CD=1.60

L =13'-0"

	Comb. 3	
	Bending	
Fb	1.500	
CD	1.150	
CF	1.000	
Cr	1.000	
CL	1.000	
Fb'	1725.00	
w D+S	57.41	pli
M	174,655	lbs-in
Sx	116.02	in 3
fb	1505.44	psi
F'b	1725.00	psi
fb	1505.44	psi
	OK	

	Comb. 8	
	Tension	
Ft	1.000	
CD	1.600	
CF	1.000	
Cr	1.000	
Ft*	1600.00	psi
ft	456.52	psi
	OK	
	Bending	
Fb	1.500	
CD	1.600	
CF	1.000	
Cr	1.000	
CL	1.000	
Fb'	2400.00	
fb	1505.44	psi
	OK	
	Combined	
ft	456.52	psi
F't	1600.00	psi
fb1	1505.44	psi
F'b1	2400.00	psi
	0.91	
	OK	

$$\frac{f'_t}{F'_t} + \frac{f'_b}{F'_b} \leq 1.0C$$

Use a 6x12 DF-L Select Structural Header Beam

Chapter 5

Design of a Roof Nailing Schedule

This chapter covers the design of roof nailing schedule.

Steps for the design of a roof nailing schedule are:

Step 1. Find the horizontal base shear from seismic analysis

Step 2. Find $R = \frac{WL}{2}$, Reaction force on the shear walls

Step 3. Find $v_{x=0} = \frac{R}{H}$, Shear flow at the boundaries of roof diaphragm

Step 4. Find $v_x = \frac{R}{H} - \frac{wx}{H}$, shear flow a distance x from the edge, where w is the uniform seismic load in plf.

Step 5. By knowing the capacity of the roof type, obtain the distances that this type is needed based on: $x = \frac{R - Hv_x}{w}$

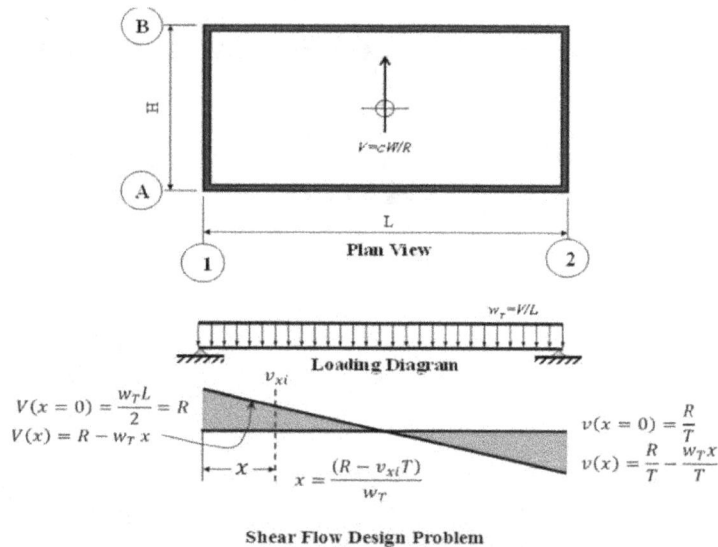

Shear Flow Design Problem

Below figure shows a typical roof nailing schedule. These sections can have different nailing types to have an optimum design.

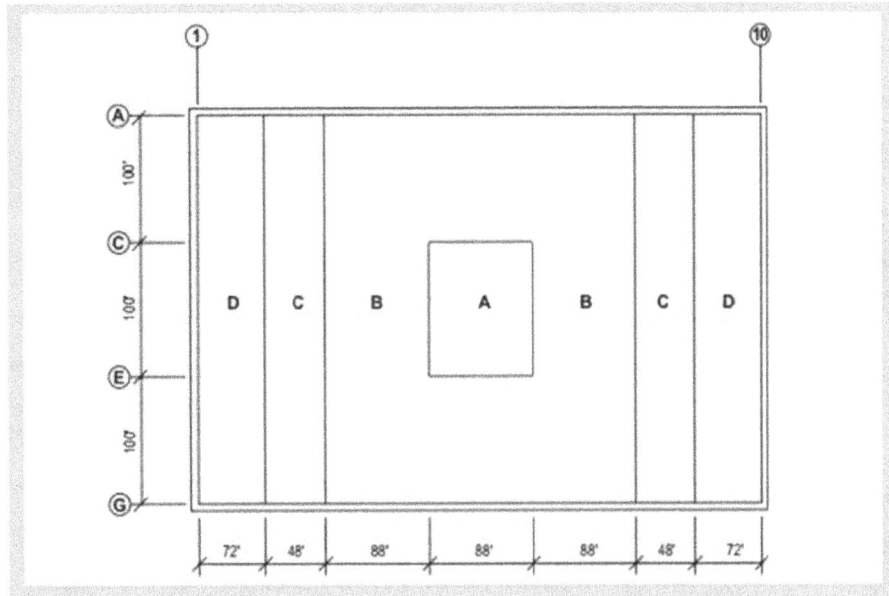

Roof Nailing Schedule:

Table 18 indicates the allowable shear for horizontal APA panel diaphragms. The example in this chapter shows how to use this table properly.

Table 18 Horizontal APA Panel Diaphragms (APA 2007)

ALLOWABLE SHEAR (POUNDS PER FOOT) FOR HORIZONTAL APA PANEL DIAPHRAGMS WITH FRAMING OF DOUGLAS-FIR, LARCH OR SOUTHERN PINE[a] FOR WIND OR SEISMIC LOADING[g] (See also IBC Table 2306.3.1)

					Blocked Diaphragms				Unblocked Diaphragms	
				Minimum Nominal Width of Framing Member at Adjoining Panel Edges and Boundaries (in.)	Nail Spacing (in.) at diaphragm boundaries (all cases), at continuous panel edges parallel to load (Cases 3 & 4), and at all panel edges (Cases 5 & 6)[b]				Nails Spaced 6" max. at Supported Edges[b]	
Panel Grade	Common Nail Size[f]	Minimum Nail Penetration in Framing (in.)	Minimum Nominal Panel Thickness (in.)		6	4	2-1/2[c]	2[c]	Case 1 (No unblocked edges or continuous joints parallel to load)	All other configurations (Cases 2, 3, 4, 5 & 6)
					Nail Spacing (in.) at other panel edges (Cases 1, 2, 3 & 4)[b]					
					6	6	4	3		
APA STRUCTURAL I grades	6d[a] (0.113" dia.)	1-1/4	5/16	2	185	250	375	420	165	125
				3	210	280	420	475	185	140
	8d (0.131" dia.)	1-3/8	3/8	2	270	360	530	600	240	180
				3	300	400	600	675	265	200
	10d[d] (0.148" dia.)	1-1/2	15/32	2	320	425	640	730	285	215
				3	360	480	720	820	320	240
APA RATED SHEATHING; APA RATED STURD-I-FLOOR and other APA grades except Species Group 5	6d[a] (0.113" dia.)	1-1/4	5/16	2	170	225	335	380	150	110
				3	190	250	380	430	170	125
			3/8	2	185	250	375	420	165	125
				3	210	280	420	475	185	140
	8d (0.131" dia.)	1-3/8	3/8	2	240	320	480	545	215	160
				3	270	360	540	610	240	180
			7/16	2	255	340	505	575	230	170
				3	285	380	570	645	255	190
			15/32	2	270	360	530	600	240	180
				3	300	400	600	675	265	200
	10d[d] (0.148" dia.)	1-1/2	15/32	2	290	385	575	655	255	190
				3	325	430	650	735	290	215
			19/32	2	320	425	640	730	285	215
				3	360	480	720	820	320	240

(a) For framing of other species: Find specific gravity for species of lumber in the AF&PA NDS. Find shear value from table above for nail size for actual grade and multiply value by the following adjustment factor: Specific Gravity Adjustment Factor = [1 − (0.5 − SG)], where SG = Specific Gravity of the framing lumber. This adjustment shall not be greater than 1.

(b) Space fasteners maximum 12 inches o.c. along intermediate framing members (6 inches o.c. when supports are spaced 48 inches o.c. or greater).

(c) Framing at adjoining panel edges shall be 3 inch nominal or wider, and nails shall be staggered where nails are spaced 2 inches o.c. or 2-1/2 inches o.c.

(d) Framing at adjoining panel edges shall be 3 inch nominal or wider, and nails shall be staggered where both of the following conditions are met: (1) 10d nails having penetration into framing of more than 1-1/2 inches and (2) nails are spaced 3 inches o.c. or less.

(e) 8d is recommended minimum for roofs due to negative pressures of high winds.

(f) The minimum nominal width of frraming members not located at boundaries or adjoining panel edges shall be 2 inches

(g) For shear loads of normal or permanent load duration as defined by AF&PA NDS, the values in the table above shall be multiplied by 0.63 and 0.56, respectively.

Note: Design for diaphragm stresses depends on direction of continuous panel joints with reference to load, not on direction of long dimension or strength axis of sheet. Continuous framing may be in either direction for blocked diaphragms.

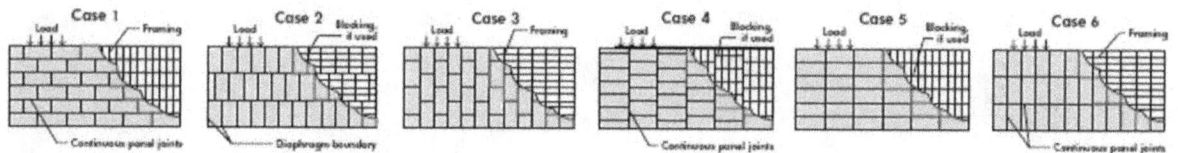

Case 1 Case 2 Case 3 Case 4 Case 5 Case 6

Different cases of framing and blocking are shown below.

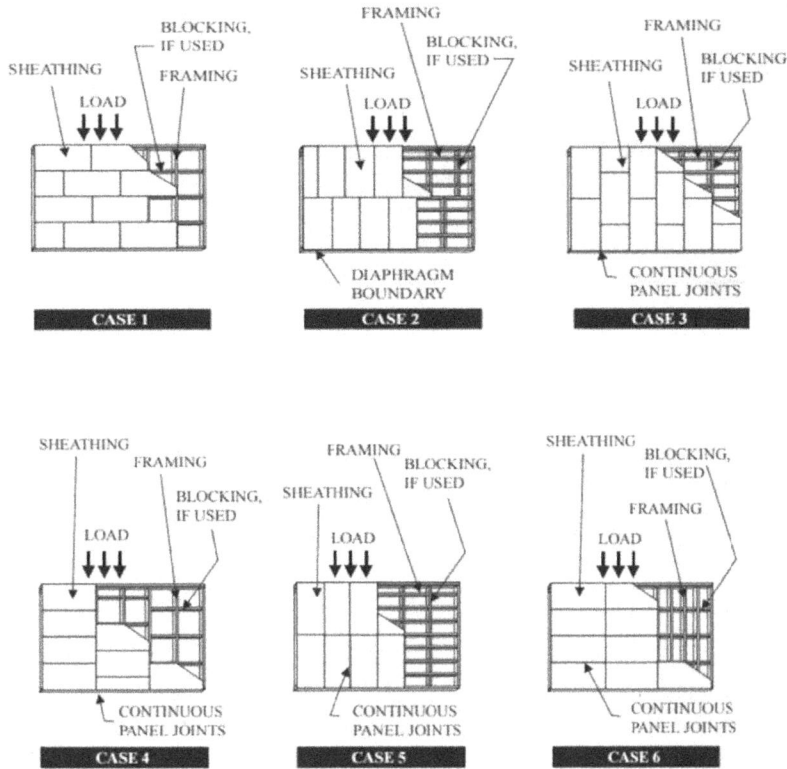

Blocki

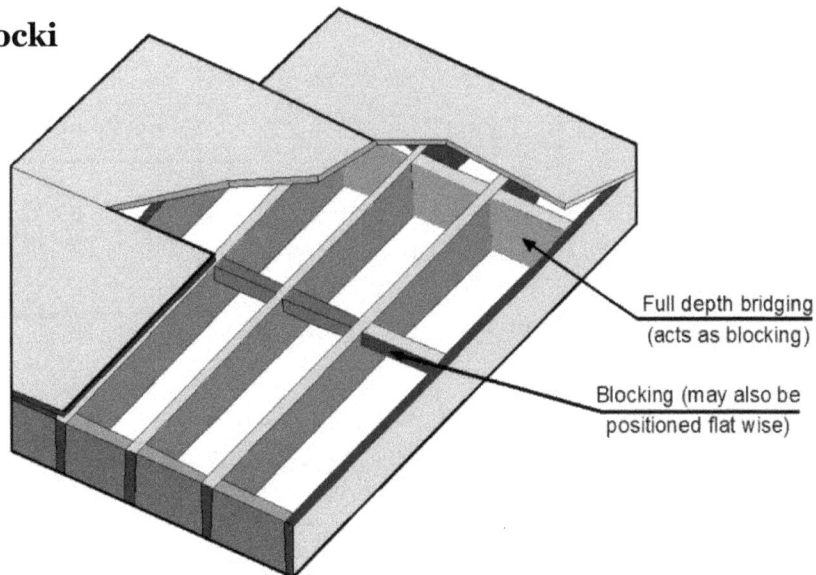

Full depth bridging
(acts as blocking)

Blocking (may also be
positioned flat wise)

Example 22:

Using the information provided in the figures below provide the roof nailing diagram design

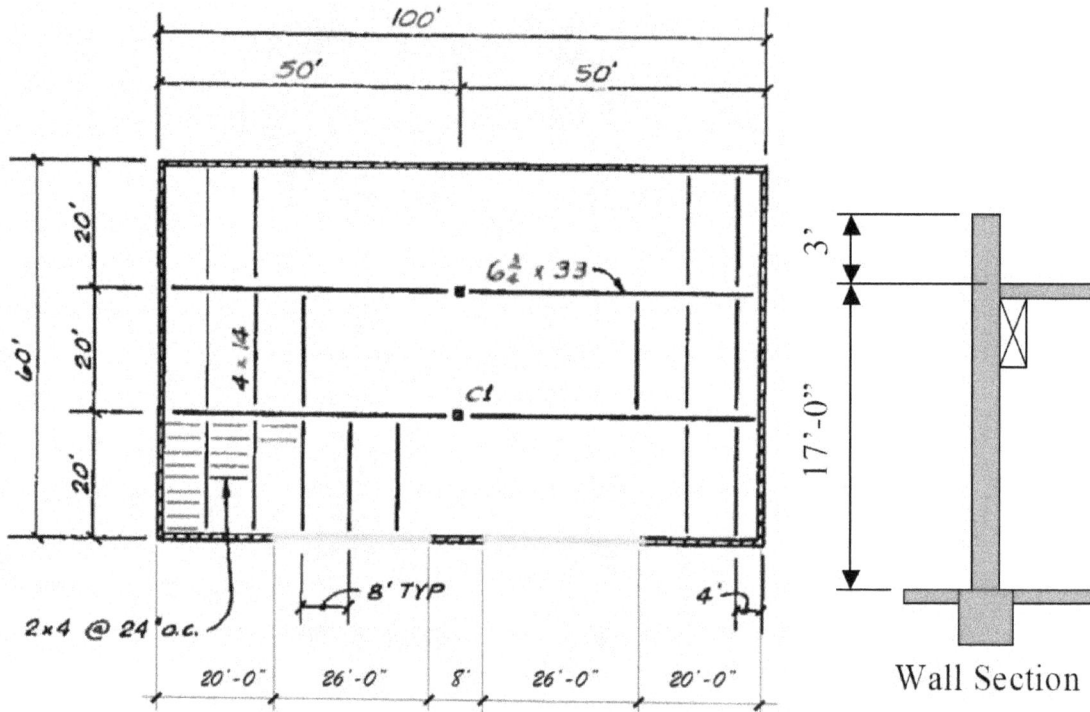

Wall Section

SEISMIC CALCULATIONS

Seismic Design Loads		
Roof L	100.00	ft
Roof W	60.00	ft
Roof A	6,000	ft2
Roof Dead	20.27	psf
Total Roof Weigth	121.60	kips
Total Wall Heigth /2	10.00	ft
Total Wall Length (perimeter)	320.00	ft
Wall Dead	15.59	
Total Wall Weigth	49.90	kips
Total Building Weigth	171.50	kips

Base Shear		
Ce	1.25	
W	171.50	kips
R	4.00	
V	53.59	kips

$$V = 1.25\frac{W}{R}$$

Calculations	E1 (l|60)	E2 (100|l)	
L	100	60	ft
v	535.95	893.25	plf
Chord Moment	1694.82	2486.05	kips-ft
Chord Distance, H	60	100	
Chord Forces (T/C)	28,247	24,860	lbs

$$M = \frac{vL^2}{8}$$

$$C = M/H$$

From example 21:

Base shear forces from seismic analysis: $V = 53.59\ kips$

Uniform shear from the seismic analysis: $v_{x=0} = 535.95\ plf$ load in E1 direction

Uniform shear from the seismic analysis: $v_{x=0} = 893.25\ plf$ load in E2 direction

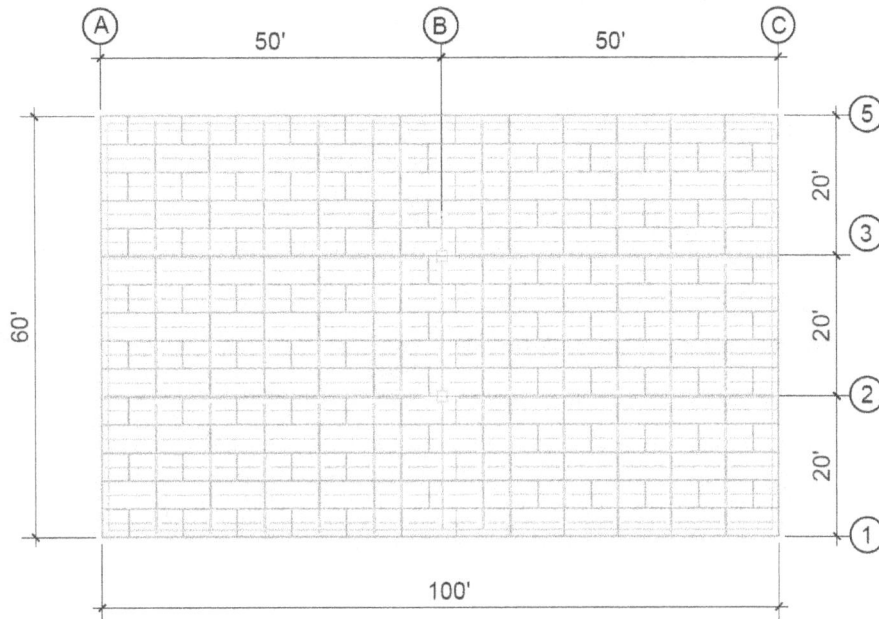

Plywood Layout

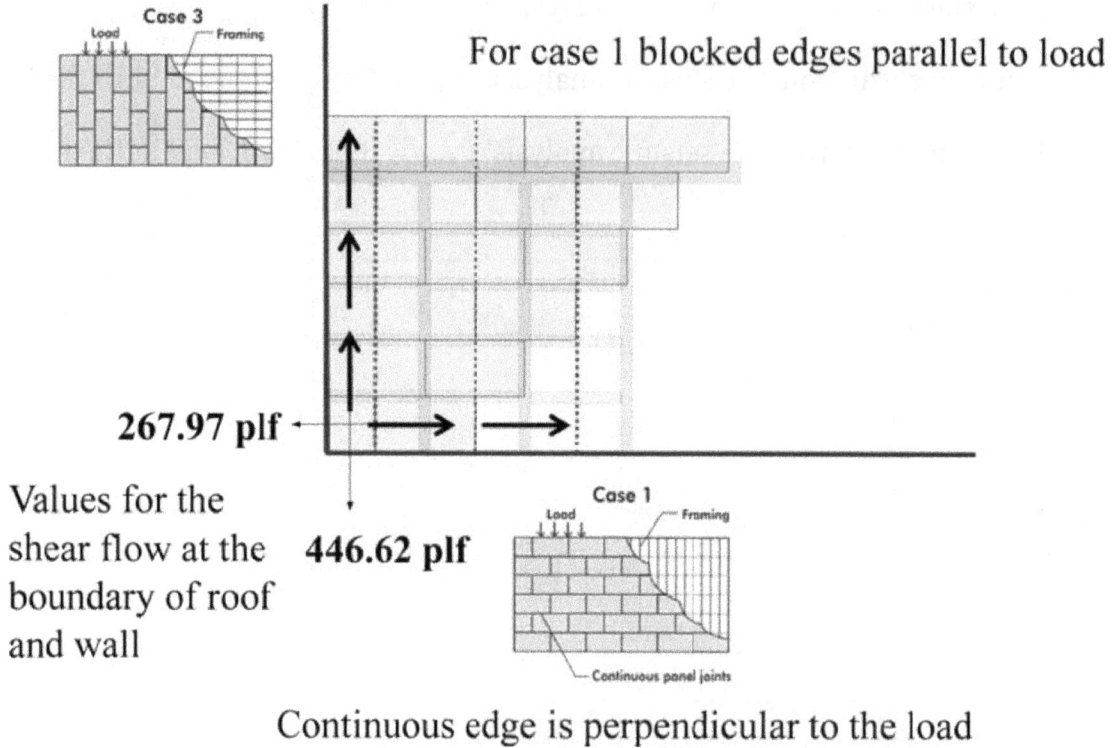

Case 3

For case 1 blocked edges parallel to load

267.97 plf

Values for the shear flow at the boundary of roof and wall

446.62 plf

Case 1

Continuous edge is perpendicular to the load

How to calculate the shear flow:

1. $R = \frac{WL}{2}$ Reaction force on the shear walls

2. $v_{x=0} = \frac{R}{H}$ Shear flow at the boundaries of roof diaphragm

3. $v_x = \frac{R}{H} - \frac{wx}{H}$ shear flow a distance x from the edge, where w is the uniform seismic load in plf

4. Knowing the capacity of the roof type, we can obtain the distances that this type is needed based on:

$$x = \frac{R - Hv_x}{w}$$

Calculations	E1 (‖60)	
L	100.00	
Seismic Load, w	535.95	
Reaction Forces, R	26.80	
H	60.00	
Shear plf	446.62	

	Capacity, vxi	APA- structu I
x0	530	4-2.5-2x
x1	360	6-4-2x
x2	270	6-6-2x
x3	240	6-0-2x

	Nailing Limits	
x0 >	0.00	
x1 <=	9.70	4-2.5-2x
x2 <=	19.77	6-4-2x
x3 <=	23.13	6-6-2x
x4 <=		6-0-2x

Calculations	E2 (100‖)	
L	60.00	
Seismic Load, w	893.25	
Reaction Forces, R	26.80	
H	100.00	
Shear plf	267.97	

	Capacity, vxi	APA- structu I
x0	270	6-6-2x
x1	180	6-0-2x

	Nailing Limits	
x0 >	0.00	
x1 <=	9.85	6-6-2x
x2 <=		6-0-2x

$$x_i = \frac{\left(R - \frac{v_{xi}H}{1000}\right)}{\frac{W}{1000}}$$

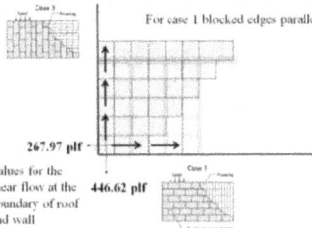

For case 1 blocked edges parallel to load

267.97 plf

446.62 plf

Values for the shear flow at the boundary of roof and wall

TABLE 2

ALLOWABLE SHEAR (POUNDS PER FOOT) FOR HORIZONTAL APA PANEL DIAPHRAGMS WITH FRAMING OF DOUGLAS-FIR, LARCH OR SOUTHERN PINE[a] FOR WIND OR SEISMIC LOADING[a] (See also IBC Table 2306.3.1)

Panel Grade	Common Nail Size[f]	Minimum Nail Penetration in Framing (in.)	Minimum Nominal Panel Thickness (in.)	Minimum Nominal Width of Framing Member at Adjoining Panel Edges and Boundaries (in.)	Blocked Diaphragms Nail Spacing (in.) at diaphragm boundaries (all cases), at continuous panel edges parallel to load (Cases 3 & 4), and at all panel edges (Cases 5 & 6)[b]				Unblocked Diaphragms Nails Spaced 6" max. at Supported Edges[b]	
					6	4	2-1/2[b]	2[b]	Case 1 (No unblocked edges or continuous joints parallel to load)	All other configurations (Cases 2, 3, 4, 5 & 6)
					Nail Spacing (in.) at other panel edges (Cases 1, 2, 3 & 4)[b]					
					6	4	3			
APA STRUCTURAL I grades	6d[a] (0.113" dia.)	1-1/4	5/16	2	185	250	375	420	165	125
				3	210	280	420	475	185	140
	8d (0.131" dia.)	1-3/8	3/8	2	270	360	530	600	240	180
				3	300	400	600	675	265	200
	10d[a] (0.148" dia.)	1-1/2	15/32	2	320	425	640	730	285	215
				3	360	480	720	820	320	240
...ATED ...HING; ...ATED ...-1-FLOOR ...her APA ...except ...s Group 5	6d[a] (0.113" dia.)	1-1/4	5/16	2	170	225	335	380	150	110
				3	190	250	380	430	170	125
			3/8	2	185	250	375	420	165	125
				3	210	280	420	475	185	140
	8d (0.131" dia.)	1-3/8	3/8	2	240	320	480	545	215	160
				3	270	360	540	610	240	180
			7/16	2	255	340	505	575	230	170
				3	285	380	570	645	255	190
			15/32	2	270	360	530	600	240	180
				3	300	400	600	675	265	200
	10d[a] (0.148" dia.)	1-1/2	15/32	2	290	385	575	655	255	190
				3	325	430	650	735	290	215
			19/32	2	320	425	640	730	285	215
				3	360	480	720	820	320	240

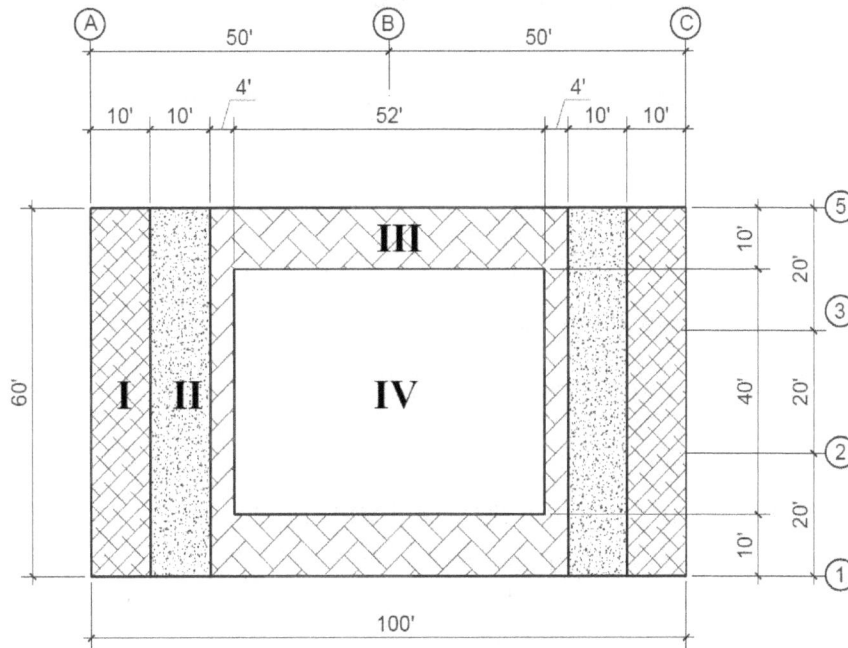

Group	Plywood, t (in.)	Minimum Framing, b	Continuous Edges (in.)	Other Edges (in.)	Blocked Edges (in.)	Capacity (plf)
I	3/8	2 x	2 ½	4	Yes	530
II			4	6	Yes	360
III			6	6	Yes	270
IV				6	No	240 (1) /180 (2)

Chapter 6

Design of a Shear Wall Nailing Schedule

This chapter shows the procedure to design shear wall nailing schedule.

Steps for design of shear wall nailing schedule

Step 1. Specify shear wall height.

Step 2. Name the shear walls in each direction.

Step 3. Find h/w ratios and calculate reduction value. (If 2<h/w<3.5, reduction value is 2w/h).

Step 4. Find total shear wall length on each side of the plan and provide reduction shear value factor.

Step 5. Find base shear and shear flow per line.

Step 6. Find modified shear flow per line.

Step 7. Use Table 20 and choose the capacity that is higher than demand (Modified shear flow).

Step 8. Use Table 20 and provide the specifications for the plywood, nail dimension, and nail spacing for each shear line.

Table 19 indicates the maximum acceptable shear wall dimension ratios. Table 20 shows the values for the capacity of different shear walls and the details for the nailing.

Example 23 covers using spreadsheet to come up with the proper design values.

Table 19 Maximum shear wall dimension ratios (Table 2305.3.4, NDS 2015)

MAXIMUM SHEAR WALL DIMENSION RATIOS	
TYPE	**MAXIMUM HEIGHT-WIDTH RATIO**
Wood structural panels or particleboard, nailed edges	For other than seismics:3 1/2:1 For seismic: 2:1(a)
Diagonal sheathing, single	2:1
Fiberboard	1 1/2:1
Gypsum board, gypsum lath, cement plaster	1 1/2:1(b)

a) For design to resist seismic forces, Shear wall height-width ratios greater than 2:1, but not exceeding 3 1/2:1 are permitted provided 2w/h multiply the allowable shear values in Table 2306.4.1 (IBC 2015).

b) Ratio shown is for unblocked construction. Height-to-width ratio is permitted to be 2:1 where the wall is installed as blocked construction in accordance with Section 2306.4.5.1.2 (IBC 2015).

Table 20 APA panel shear walls capacity (APA 2007)

ALLOWABLE SHEAR (POUNDS PER FOOT) FOR APA PANEL SHEAR WALLS WITH FRAMING OF DOUGLAS-FIR, LARCH, OR SOUTHERN PINE(a) FOR WIND OR SEISMIC LOADING(b,h,i,j,k) (See also IBC Table 2306.4.1)

Panel Grade	Minimum Nominal Panel Thickness (in.)	Minimum Nail Penetration in Framing (in.)	Panels Applied Direct to Framing					Panels Applied Over 1/2" or 5/8" Gypsum Sheathing				
			Nail Size (common or galvanized box)(k)	Nail Spacing at Panel Edges (in.)				Nail Size (common or galvanized box)	Nail Spacing at Panel Edges (in.)			
				6	4	3	2(e)		6	4	3	2(e)
APA STRUCTURAL I grades	5/16	1-1/4	6d (0.113" dia.)	200	300	390	510	8d (0.131" dia.)	200	300	390	510
	3/8	1-3/8	8d (0.131" dia.)	230(d)	360(d)	460(d)	610(d)	10d (0.148" dia.)	280	430	550(f)	730
	7/16			255(d)	395(d)	505(d)	670(d)					
	15/32			280	430	550	730					
	15/32	1-1/2	10d (0.148" dia.)	340	510	665(f)	870		—	—	—	—
APA RATED SHEATHING; APA RATED SIDING(g) and other APA grades except Species Group 5	5/16 or 1/4(c)	1-1/4	6d (0.113" dia.)	180	270	350	450	8d (0.131 dia.)	180	270	350	450
	3/8			200	300	390	510		200	300	390	510
	3/8		8d (0.131" dia.)	220(d)	320(d)	410(d)	530(d)	10d (0.148" dia.)	260	380	490(f)	640
	7/16	1-3/8		240(d)	350(d)	450(d)	585(d)					
	15/32			260	380	490	640					
	15/32	1-1/2	10d (0.148" dia.)	310	460	600(f)	770		—			
	19/32			340	510	665(f)	870		—			
APA RATED SIDING(g) and other APA grades except Species Group 5			Nail Size (galvanized casing)					Nail Size (galvanized casing)				
	5/16(c)	1-1/4	6d (0.113" dia.)	140	210	275	360	8d (0.131" dia.)	140	210	275	360
	3/8	1-3/8	8d (0.131" dia.)	160	240	310	410	10d (0.148" dia.)	160	240	310(f)	410

(a) For framing of other species: Find specific gravity for species of lumber in the AF&PA National Design Specification (NDS). Find shear value from table above for nail size for actual grade and multiply value by the following adjustment factor: Specific Gravity Adjustment Factor = [1 – (0.5 – SG)], where SG = Specific Gravity of the framing lumber. This adjustment shall not be greater than 1.

(b) Panel edges backed with 2 inch nominal or wider framing. Install panels either horizontally or vertically. Space fasteners maximum 6 inches on center along intermediate framing members for 3/8 inch and 7/16 inch panels installed on studs spaced 24 inches on center. For other conditions and panel thicknesses, space nails maximum 12 inches on center on intermediate supports.

(c) 3/8 inch panel thickness or siding with a span rating of 16 inches on center is the minimum recommended where applied direct to framing as exterior siding.

(d) Allowable shear values are permitted to be increased to values shown for 15/32 inch sheathing with same nailing provided (1) studs are spaced a maximum of 16 inch on center, or (2) panels are applied with long dimension across studs.

(e) Framing at adjoining panel edges shall be 3 inch nominal or wider, and nails shall be staggered where nails are spaced 2 inch on center.

(f) Framing at adjoining panel edges shall be 3 inch nominal or wider, and nails shall be staggered where both the following conditions are met: (1) 10d (3 inch x 0.148 inch) nails having penetration into framing of more than 1-1/2 inch and (2) nails are spaced 3 inch on center.

(g) Values apply to all-veneer plywood. Thickness at point of fastening on panel edges governs shear values.

(h) Where panels applied on both faces of a wall and nail spacing is less than 6 inches o.c. on either side, panel joints shall be offset to fall on different framing members, or framing shall be 3 inch nominal or thicker at adjoining panel edges and nails on each side shall be staggered.

(i) In Seismic Design Category D, E or F, where shear design values exceed 350 pounds per lineal foot, all framing members receiving edge nailing from abutting panels shall not be less than a single 3 inch nominal member, or two 2 inch nominal members fastened together in accordance with IBC Section 2306.1 to transfer the design shear value between framing members. Wood structural panel joint and sill plate nailing shall be staggered in all cases. See IBC Section 2305.3.11 for sill plate size and anchorage requirements.

(j) Galvanized nails shall be hot dipped or tumbled.

(k) For shear loads of normal or permanent load duration as defined by the AF&PA NDS, the values in the table above shall be multiplied by 0.63 or 0.56, respectively.

Typical Layout for Shear Walls

Example 23:

Using the information provided in the figures below provide the shear walls nailing and plywood.

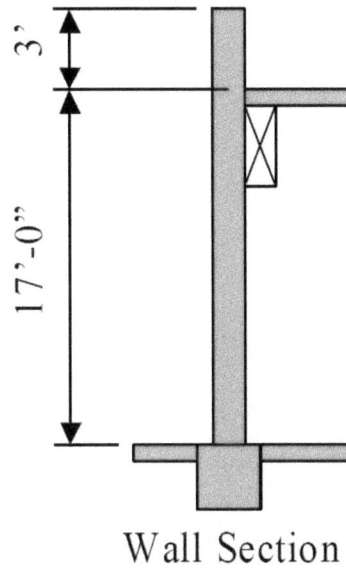

Wall Section

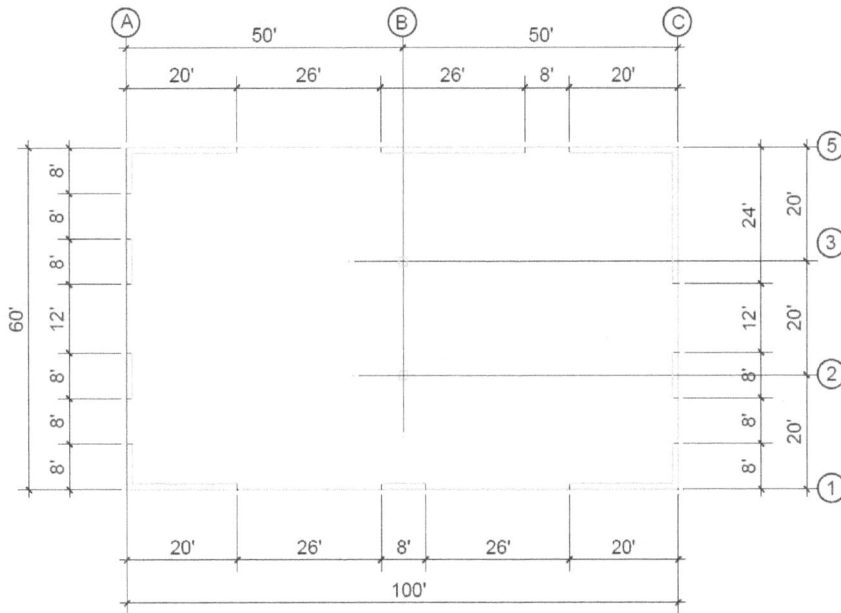

Calculations and how to use Table 19 and Table 20:

Shear Wall Heigth, h	16.5				
Shear Wall Along Line A					
Wall No	1	2	3	4	
Shear Wall Length, w	8.00	8.00	8.00	8.00	Reduce Value
h/w ratio	2.06	2.06	2.06	2.06	0.97
Reduce Shear Value	0.97	0.97	0.97	0.97	Total L
Total Lsw	8.00	8.00	8.00	8.00	32.00
Shear Wall Along Line B					
Wall No	1	2	3		
Shear Wall Length, w	8.00	8.00	24.00		Reduce Value
h/w ratio	2.06	2.06	0.69		0.97
Reduce Shear Value	0.97	0.97	1.00		Total L
Total Lsw	8.00	8.00	24.00		40.00
Shear Wall Along Line 1					
Wall No	1	2	3		
Shear Wall Length, w	20.00	8.00	20.00		Reduce Value
h/w ratio	0.83	2.06	0.83		0.97
Reduce Shear Value	1.00	0.97	1.00		Total L
Total Lsw	20.00	8.00	20.00		48.00
Shear Wall Along Line 5					
Wall No	1	2	3		
Shear Wall Length, w	20.00	26.00	20.00		Reduce Value
h/w ratio	0.83	0.63	0.83		1.00
Reduce Shear Value	1.00	1.00	1.00		Total L
Total Lsw	20.00	26.00	20.00		66.00
Calculations					
Shear Base Force	53.59	53.59	53.59	53.59	kips
Shear Wall Reaction Forces	26.80	26.80	26.80	26.80	kips
Shear Walls Line	A	B	1	5	
Wall L	32.00	40.00	48.00	66.00	ft
Reduced Shear Ratio	0.97	0.97	0.97	1.00	
Shear Flow	837.42	669.94	558.28	406.02	plf
Modified Shear Flow	863.59	690.87	575.73	406.02	plf
Plywood Type	15/32	15/32	15/32	15/32	
Nail	10d	8d	10d	8d	
Nail Space	2	2	3	4	
Capacity	870	730	665	430	plf
	OK	OK	OK	OK	

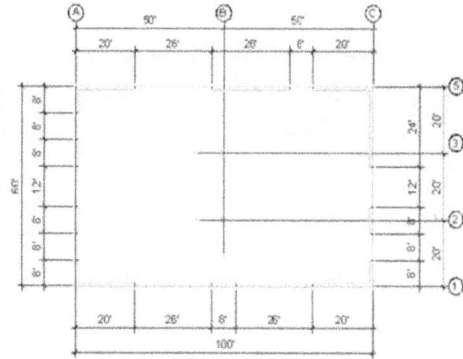

ALLOWABLE SHEAR (POUNDS PER FOOT) FOR APA PANEL SHEAR WALLS WITH FRAMING OF DOUGLAS-FIR, LARCH, OR SOUTHERN PINE[a] FOR WIND OR SEISMIC LOADING[b,c,d] (See also IBC Table 2306.4.1)

Panel Grade	Minimum Nominal Panel Thickness (in.)	Minimum Nail Penetration in Framing (in.)	Panels Applied Direct to Framing					Panels Applied Over 1/2" or 5/8" Gypsum Sheathing				
			Nail Size (common or galvanized box)[h]	Nail Spacing at Panel Edges (in.)				Nail Size (common or galvanized box)	Nail Spacing at Panel Edges (in.)			
				6	4	3	2[g]		6	4	3	2[g]
APA STRUCTURAL I grades	5/16	1-1/4	6d (0.113" dia.)	200	300	390	510	8d (0.131" dia.)	200	300	390	510
	3/8		8d (0.131" dia.)	230[e]	360[e]	460[e]	610[e]	10d (0.148" dia.)	280	430	550[e]	730
	7/16	1-3/8		255[e]	395[e]	505[e]	670[e]					
	15/32			280	430	550	730					
	15/32	1-1/2	10d (0.148" dia.)	340	510	665[e]	870		—	—	—	—
APA RATED SHEATHING; APA RATED SIDING[f] and other APA grades except Species Group 5	5/16 or 1/4[e]	1-1/4	6d (0.113" dia.)	180	270	350	450	8d (0.131" dia.)	180	270	350	450
	3/8			200	300	390	510		200	300	390	510
	3/8		8d (0.131" dia.)	220[e]	320[e]	410[e]	530[e]	10d (0.148" dia.)	260	380	490[e]	640
	7/16	1-3/8		240[e]	350[e]	450[e]	585[e]					
	15/32			260	380	490	640					
	15/32	1-1/2	10d (0.148" dia.)	310	460	600[e]	770		—	—	—	—
	19/32			340	510	665[e]	870					

References

The following references were used to create this book and can be utilized as a reference for analysis and design of wood structures.

AASHTO LRFD Bridge Design Specifications, 7th Edition with 2015 Interim Revisions. Published by the American Association of State Highway and Transportation Officials (AASHTO), Washington, D.C.

American Forest & Paper Association (AF&PA). ASD/LRFD National Design Specification for Wood Construction with Commentary and Supplement: Design Values for Wood Construction. Washington, DC: AF&PA American Wood Council, 2015.

American Institute of Timber Construction (AITC). Timber Construction Manual. 5th ed. New York: John Wiley & Sons, 2005.

American Plywood Association (APA). Design/Construction Guide: Diaphragms and Shear Walls. Tacoma, WA: APA-The Engineered Wood Association, 2007.

American Plywood Association (APA), Panel Design Specification. Tacoma, WA: APA The Engineered Wood Association, 2008.

American Plywood Association (APA), Plywood Design Specification. Tacoma, WA: APA-The Engineered Wood Association, 1998.

American Plywood Association (APA), Plywood Design Specification, Supplement 2: Design and Fabrication of Plywood-Lumber Beams. Tacoma, WA: APA-The Engineered Wood Association, Tacoma, Wash, 1992.

American Plywood Association (APA)–The Engineered Wood Association. "Diaphragms and shear walls: Design/construction guide." APA–The Engineered Wood Association, Tacoma, Wash, 2007.

American Society of Civil Engineers (ASCE), Minimum Design Loads for Buildings and Other Structures. New York: ASCE, 2006.

Breyer, Donald E., Kelly E. Cobeen, Kenneth J. Fridley, and David G. Pollock, Design of Wood Structures- ASD. 5th ed. New York: McGraw-Hill, 2007.

International Building Code. Country Club Hills, Ill: ICC, 2015. Print.

Kim, Jai B., and Robert H. Kim, "Oak A-Frame Timber Bridges Meeting the Modern Deflection Requirement," Transportation Research Record No. 1319, Washington DC, 1991.

"Simpson Wood Construction Connectors Catalog 2017." <https://www.strongtie.com/resources/literature/wood-construction-connectors-catalog>.

U.S. Department of Agriculture, Forest Products Laboratory. Wood Handbook: Wood as an Engineering Material. Honolulu, Ill: University Press of the Pacific, 2006.

Appendix – Design Project

In this project, an application of using chapters of this book is presented. Design a two-story building with the following plans.

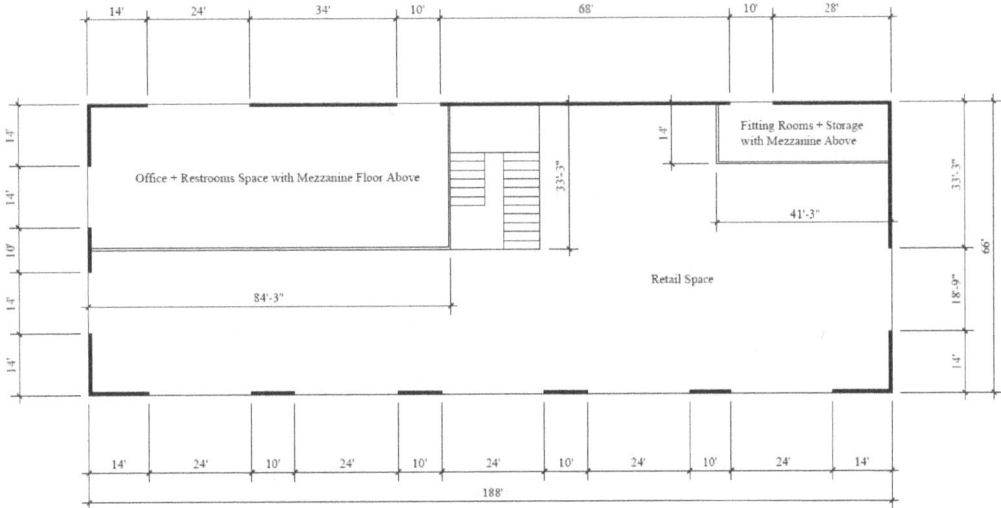

First Floor Plan

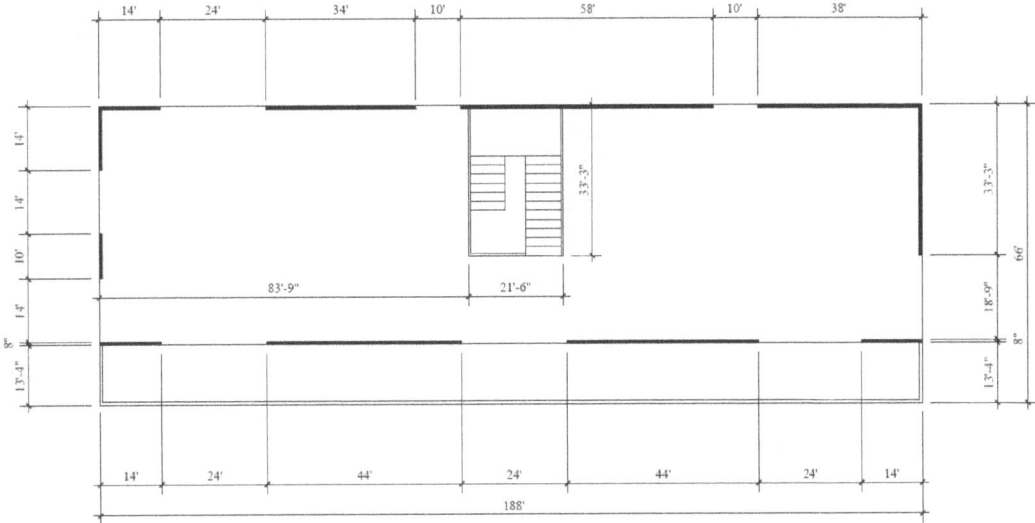

Office + Restrooms Space with Mezzanine Floor Above

Fitting Rooms + Storage with Mezzanine Above

Retail Space

84'-3"

41'-3"

33'-3"

18'-9"

66'

188'

14' | 24' | 34' | 10' | 68' | 10' | 28'

14' | 24' | 10' | 24' | 10' | 24' | 10' | 24' | 10' | 24' | 14'

14' | 14' | 10' | 14' | 14'

Second Floor Plan

83'-9"

21'-6"

33'-3"

18'-9"

66'

188'

14' | 24' | 34' | 10' | 58' | 10' | 38'

14' | 24' | 44' | 24' | 44' | 24' | 14'

14' | 14' | 10' | 14' | 8'

13'-4"

8"

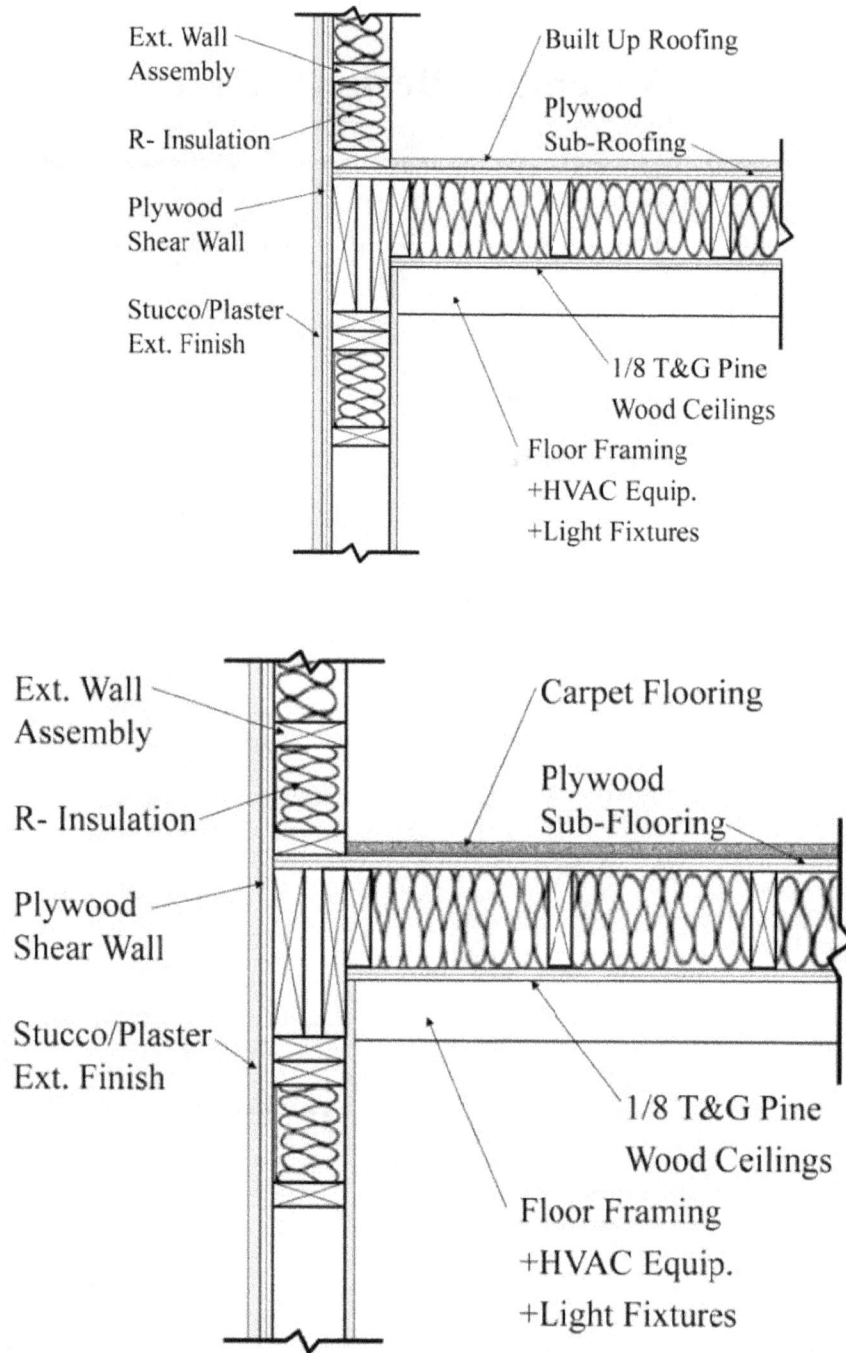

Floor and roof cross sections

A sample of Architectural details:

Design Project

The Design covers following sections:

Section A: Project Statement

Section B: Building Description

Section C: Framing Plans - Building

Section D: Gravity and Lateral Loads - Building

Section E: Glulam Beam Design - Floor

Section F: Glulam Beam Design - Roof

Section G: Column Design - Floor

Section H: Column Design - Roof

Section I: Purlin Design - Floor

Section J: Purlin Design - Roof

Section K: Sub-purlin Design - Floor

Section L: Sub-purlin Design - Roof

Section M: Header Design - Floor

Section N: Header Design - Roof

Section O: Front Balcony Design Alternative 1 - Floor

Section P: Front Balcony Design Alternative 2 - Floor

Section Q: Nailing Schedule - Floor

Section R: Nailing Schedule – Roof

Section S: Shear Wall nailing Design - Floor

Section T: Shear Wall nailing Design - Roof

Design Project Statement:

Provide a complete structural design for given framing sections shown in the next few pages. The snow load is 65 psf. This load has already been converted to a flat roof analogy. The floor live load is 50 psf.

General Requirements:

A. Material Conditions for the Glu-Lam Beams

- Use a 24F-V4 SP/SP, southern species.
- Use 1 ½ " thick laminations.
- Lu design bracing as required.
- MC < 16%.
- Beams supports wood finish with the same design conditions for deflection as plaster finish.

B. Material Conditions for the Purlins:

- May use a 2 1/8" glu-lam joists 24F-V4 SP/SP, southern species
- Use 1 ½ " thick laminations.
- Lu design bracing as required.
- MC < 16%.
- Beams supports wood finish with the same design conditions for deflection as plaster finish.

C. Material Conditions for the Purlins:

- Must use Douglas Fir-larch Select Structural Members.
- Lu design bracing as required.
- MC < 19%.
- Beams supports wood finish with the same design conditions for deflection as plaster finish.

D. Material Conditions for the Columns:

- Use a square section.
- Douglas Fir-larch Select Structural Members.
- MC < 15%.

E. Material Conditions for the Header Beams:

- Use a 6x member.
- Douglas Fir-larch Select Structural Members
- Lu design bracing as required. Members framing into the header beam can be considered braces for the Header
- MC < 15%.

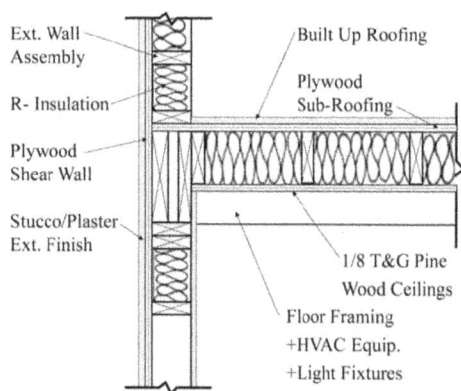

Roof Cross Section

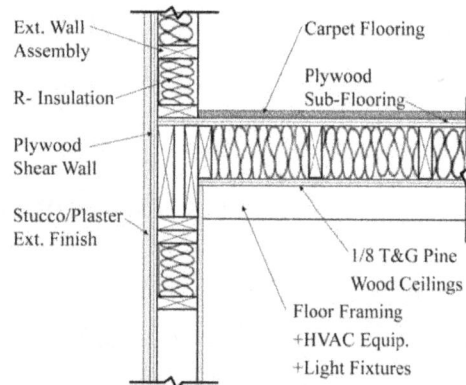

Floor Cross Section

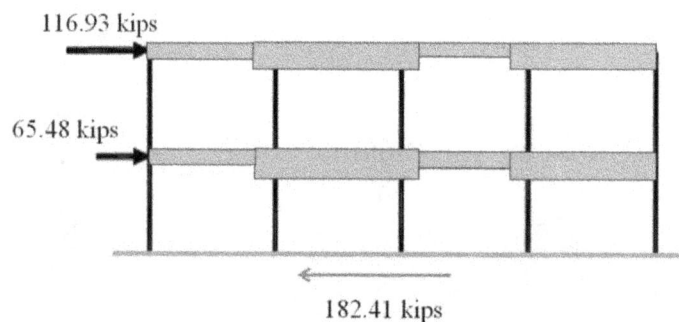

Seismic Forces

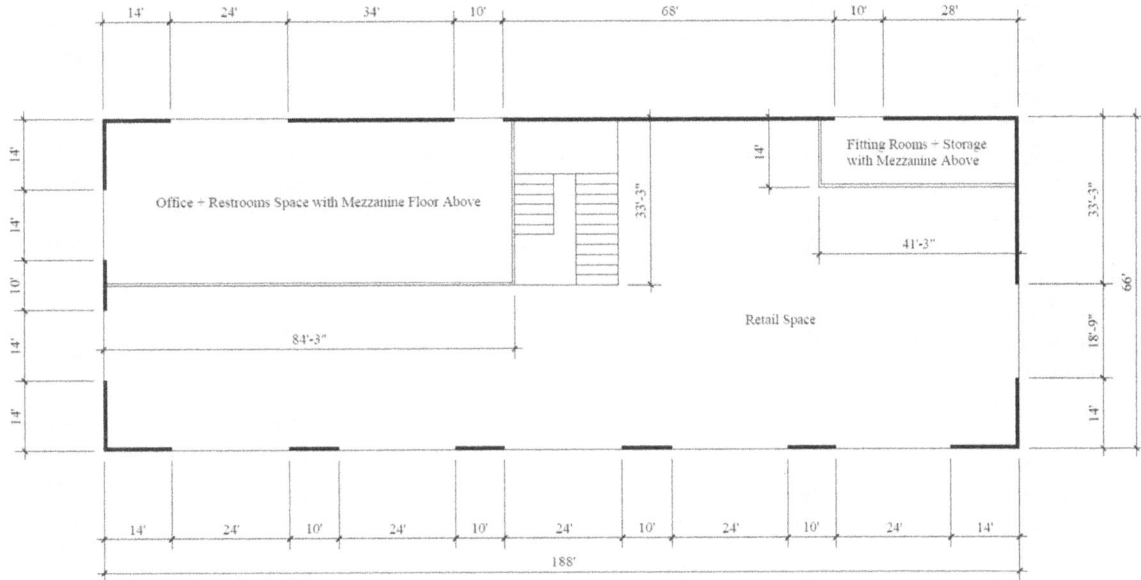

First Floor Plan

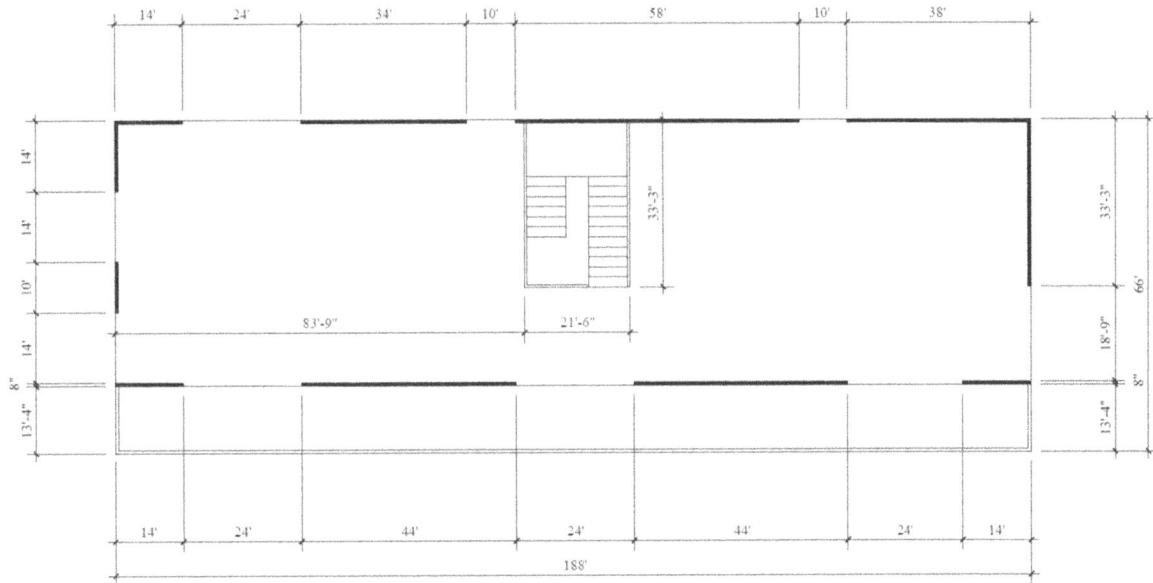

Second Floor Plan

Introduction and description:

The Purpose of this project is designing a two-story Wood Structure Building in accordance with IBC and NSD by using ASD method.

The Main goal of this project is having a durable and cost effective wood structure and guidelines for the builder. Therefore, all of the building elements are designed and analyzed except the foundation. The Design philosophy is based on ASD (Allowable Stress Design). Thus, all of the load combinations related to this type of designs are considered. It is important to mention that wood structures experience high degrees of indeterminacy. To be conservative and ease the calculations, designs may be performed for simple statically structures. Moreover, For meeting this criterion the main beams are also statically determinate structures with having hinges at a certain location.

The wood members and types of woods are selected based on the popular and most available materials in the market. The connections are designed for connecting different members, and The SIMPSON products are utilized for connection design.

According to the given plan, Layout of plywood, Purlin, Sub-purlin, Glu-Lam Beam (GLB), and locations of hinges are determined to have the most efficient use of plywood by having a very small loss. The optimization process is performed based on the location of hinges to have the least volume of wood.

Building Occupancy:

This Building is a two story Multi-use building in the area with high intensity of snow.

Area of Base Floor: 12,408 ft^2

Area of First Floor: 12,408 ft^2

Area of Second Floor: 9,776 ft^2

Total Area: 34,592 ft^2

It can be noticed that this building will deliver a decent area for its occupancy.

The occupancy of Base floor is a retail Store and Ski Resort, the first and second floor are office spaces. It is important to mention that the balcony located on the second floor also considered having a live load from office space. In other words, there are some posts and walls on the perimeter of the balcony to avoid entering snow to the balcony's floor. Dead Load is calculated based on floor detailing in each floor in addition to self-weight of members on that particular floor. If it is not mentioned the density of wood is assumed 36 pcf.

Here is the list of used code for performing analysis and design for this project.

1. International Building Code (IBC)
2. American Wood Council (ASD)
3. MINIMUM DESIGN LOADS FOR BUILDINGS AND OTHER STRUCTURES ASCE 07

Building Manual and catalogs Used:
1. National Design wood specifications (2015)
2. Simpson Strong Ties Connectors (2017)

FRAMING PLANS - BUILDING

GLBs

Columns

Purlins

Sub-Purlins

Headers

Shear Walls

Ply-Wood Layout

Balcony

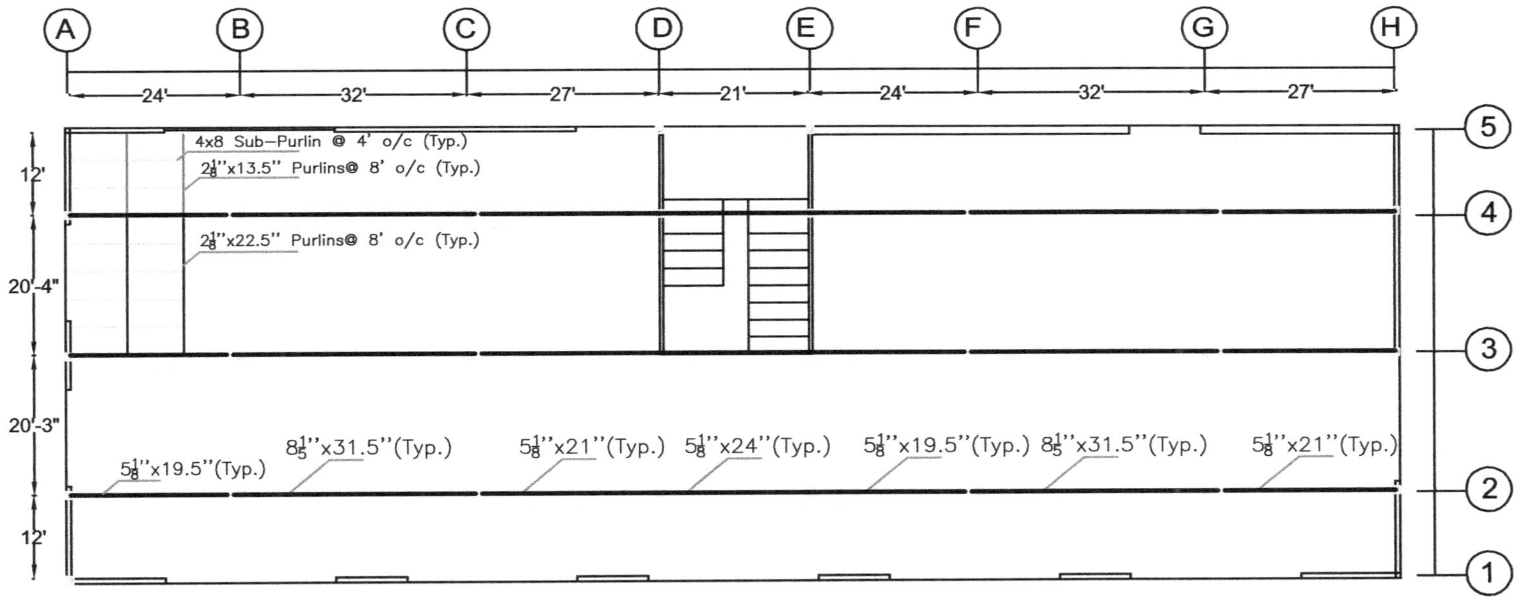

A
B
C
D
E
F
G
H
24'
32'
27'
21'
24'
32'
27'
1
2
3
4
5
12'
20'-4"
20'-3"
12'
4x8 Sub–Purlin @ 4' o/c (Typ.)
2$\frac{1}{8}$"x13.5" Purlins@ 8' o/c (Typ.)
2$\frac{1}{8}$"x22.5" Purlins@ 8' o/c (Typ.)
5$\frac{1}{8}$"x19.5"(Typ.)
8$\frac{1}{5}$"x31.5"(Typ.)
5$\frac{1}{8}$"x21"(Typ.)
5$\frac{1}{8}$"x24"(Typ.)
5$\frac{1}{8}$"x19.5"(Typ.)
8$\frac{1}{5}$"x31.5"(Typ.)
5$\frac{1}{8}$"x21"(Typ.)

First Floor

REFERENCE

REFERENCE. TITLE | REFERENCE. No.

NOTES

Main–Plan
Floor

KEY PLAN

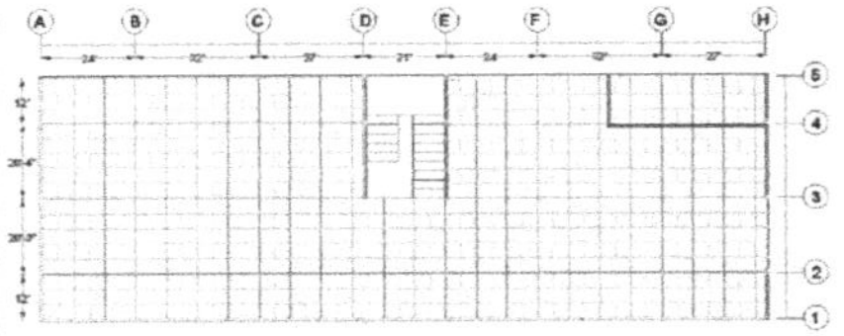

First Floor

EQUIPMENT LIST

ABBREVIATIONS

ARZHANG ZAMANI

Design of Wood Structures

DOCUMENT | CONTRACT | DATE | A1 | SCALE | X OF Y | REV.

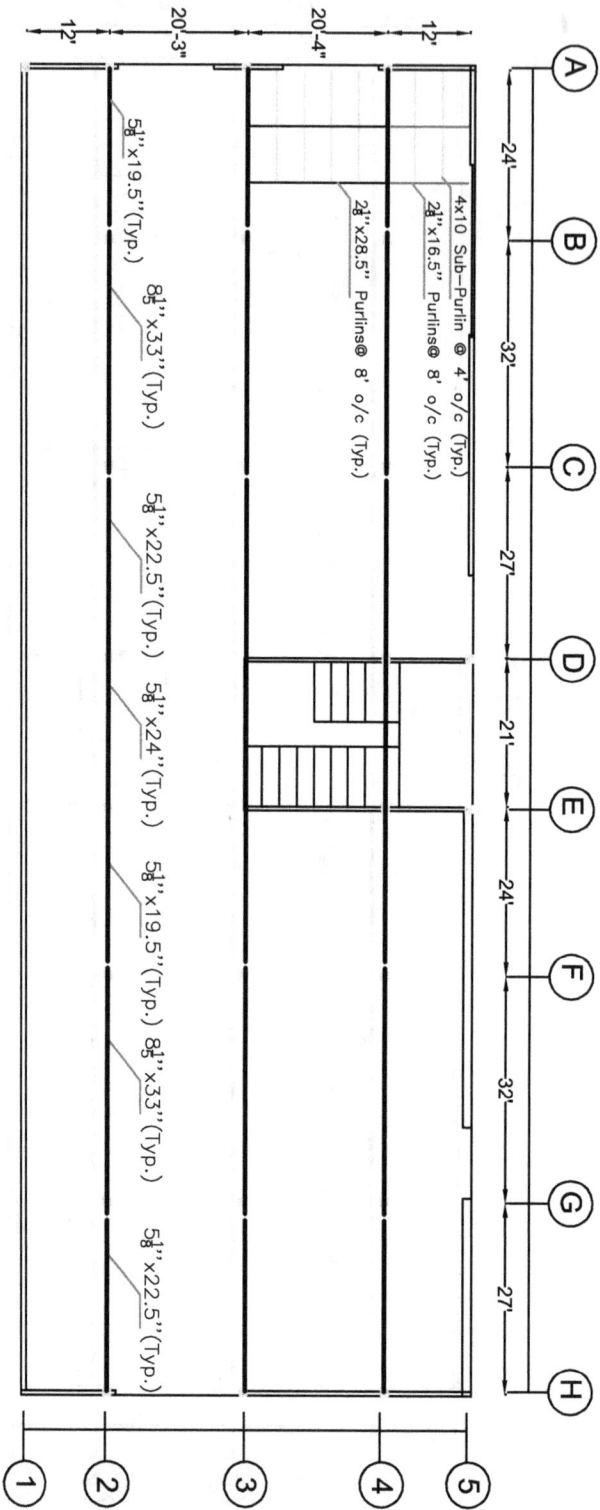

Roof

4x10 Sub-Purlin @ 4" o/c (Typ.)

$2\frac{1}{8}$"x16.5" Purlins@ 8' o/c (Typ.)

$2\frac{1}{8}$"x28.5" Purlins@ 8' o/c (Typ.)

$5\frac{1}{8}$"x19.5"(Typ.)

$8\frac{1}{8}$"x33"(Typ.)

$5\frac{1}{8}$"x22.5"(Typ.)

$5\frac{1}{8}$"x24"(Typ.)

$5\frac{1}{8}$"x19.5"(Typ.)

$8\frac{1}{8}$"x33"(Typ.)

$5\frac{1}{8}$"x22.5"(Typ.)

12'   20'-3"   20'-4"   12'

24'   32'   27'   21'   24'   32'   27'

A B C D E F G H

1 2 3 4 5

Main Plan
Roof

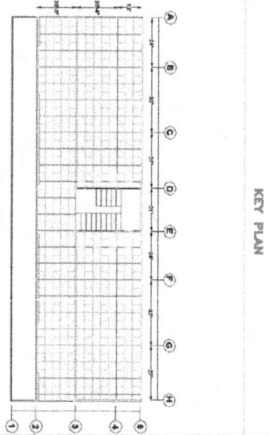

KEY PLAN

Second Floor

EQUIPMENT LIST

ABBREVIATIONS

ISSUED FOR REVIEW/COMMENT
DESCRIPTION

PROJECT:

CLIENT:

Design of Wood Structures

Title:

DOCUMENT No.   CONTRACT No.   DATE
DOCUMENT   CONTRACT   DATE   ORIGINAL SIZE   SCALE   SHEET No.   REV.
14   15   16   A1   SCALE   X OF Y   REV.

ARZHANG ZAMANI

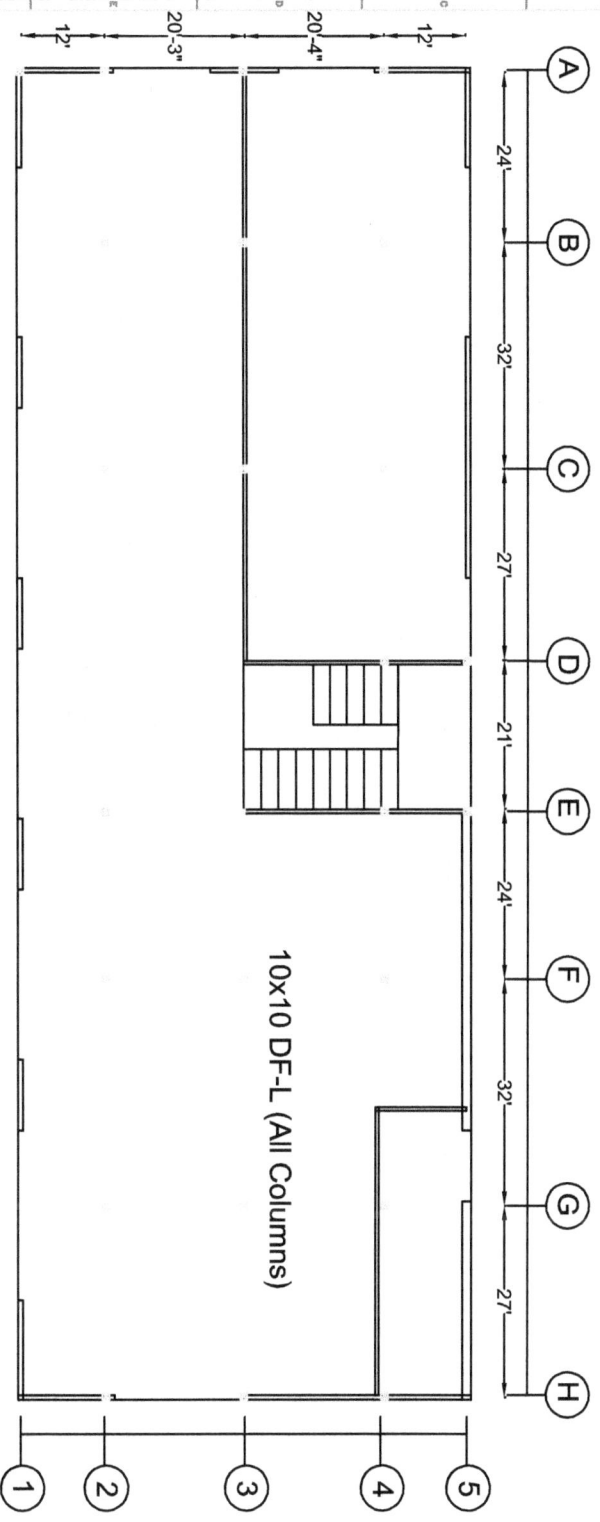

First Floor

10x10 DF-L (All Columns)

Columns: A B C D E F G H

24' 32' 27' 21' 24' 32' 27'

Rows: 1 2 3 4 5

12' 20'-4" 20'-3" 12'

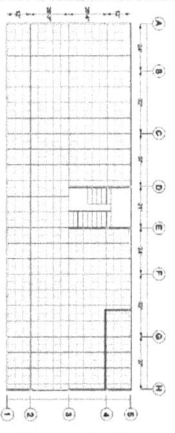

Design of Wood Structures

AZZHANG ZAMANI

PREPARED

CLIENT

REFERENCE, TITLE   REFERENCE   REFERENCE, No.

NOTES

KEY PLAN

Columns
Floor

First Floor

EQUIPMENT LIST

ABBREVIATIONS

ISSUED FOR DEVELOPMENT

DOCUMENT No.   CURRENT No.   DRAWN   CHECKED   APPROVED
DOCUMENT   CONTRACT   DATE   DATE
14   15   16

ORIGINAL SIZE   SCALE   SHEET No.   REV
A1   X   OF   Y   REV

07 Jul 2008

Title

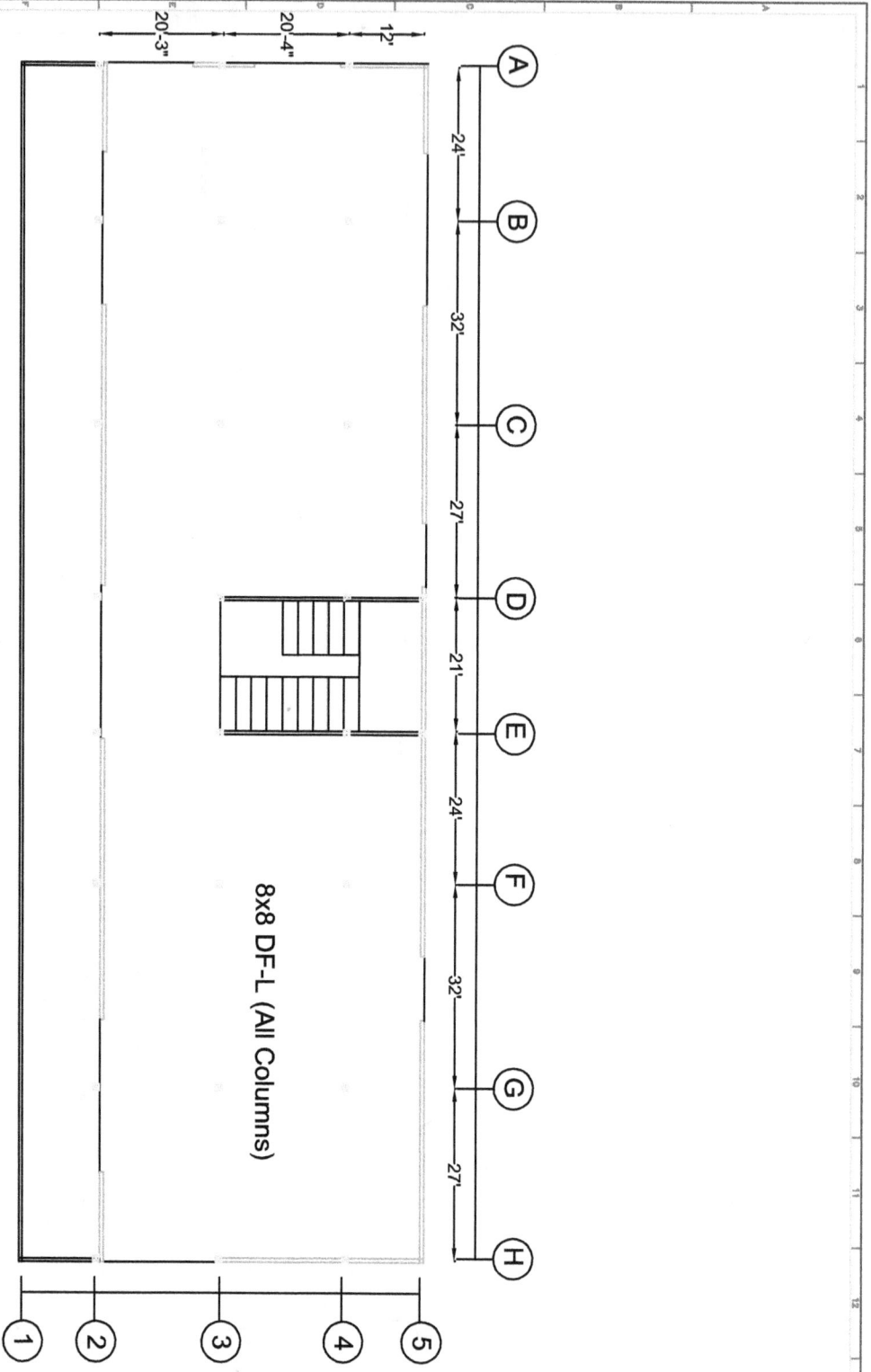

Second Floor

8x8 DF-L (All Columns)

Columns
Roof

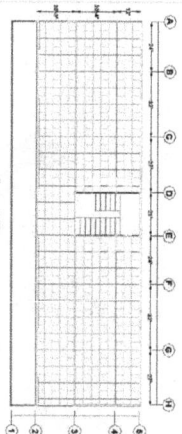

Design of Wood Structures

First Floor

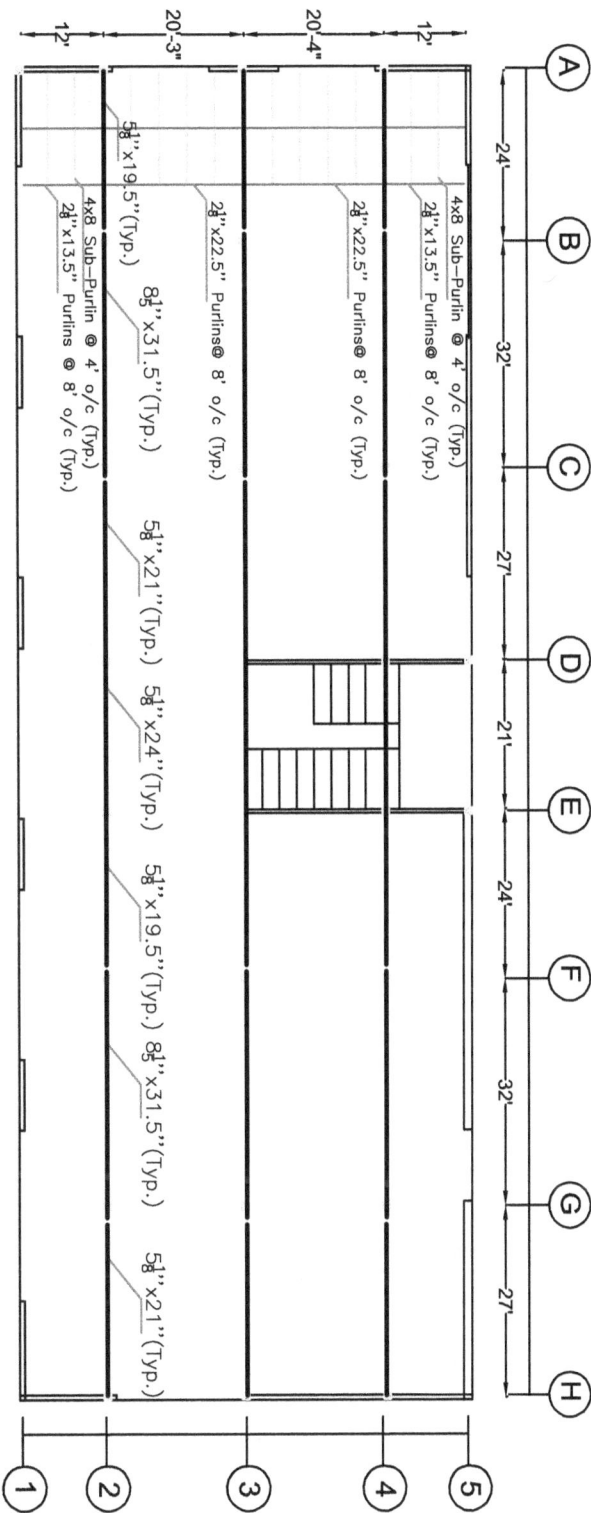

Grid columns (top): A B C D E F G H

Horizontal dimensions: 24' 32' 27' 21' 24' 32' 27'

Grid rows (bottom): 1 2 3 4 5

Vertical dimensions: 12' 20'-4" 20'-3" 12'

Labels:
- 4x8 Sub-Purlin @ 4' o/c (Typ.)
- $2\frac{1}{4}$"x13.5" Purlins@ 8' o/c (Typ.)
- $2\frac{1}{4}$"x22.5" Purlins@ 8' o/c (Typ.)
- $2\frac{1}{4}$"x22.5" Purlins @ 8' o/c (Typ.)
- $5\frac{1}{8}$"x19.5"(Typ.)
- $8\frac{1}{2}$"x31.5"(Typ.)
- $5\frac{1}{8}$"x21"(Typ.)
- $5\frac{1}{8}$"x24"(Typ.)
- $5\frac{1}{8}$"x19.5"(Typ.)
- $8\frac{1}{2}$"x31.5"(Typ.)
- $5\frac{1}{8}$"x21"(Typ.)
- 4x8 Sub-Purlin @ 4' o/c (Typ.)
- $2\frac{1}{4}$"x13.5" Purlins @ 8' o/c (Typ.)

Purlin and Sub-Purlin Floor

REFERENCE, TITLE   NOTES   REFERENCE

REFERENCE, No.

KEY PLAN

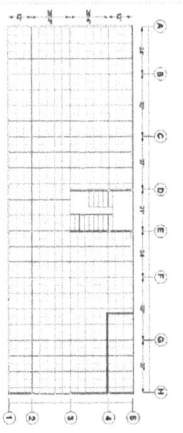

First Floor

EQUIPMENT LIST   ABBREVIATIONS

DESIGN FOR CONSTRUCTION

DOCUMENT NO.   CONTRACT   DATE

ARCHANG ZAMANI

Design of Wood Structures

DRAWN   CHECKED   APPROVED

CONTROL REF. SCALE SHEET NO. REV.

A1   SCALE   X OF Y   REV.

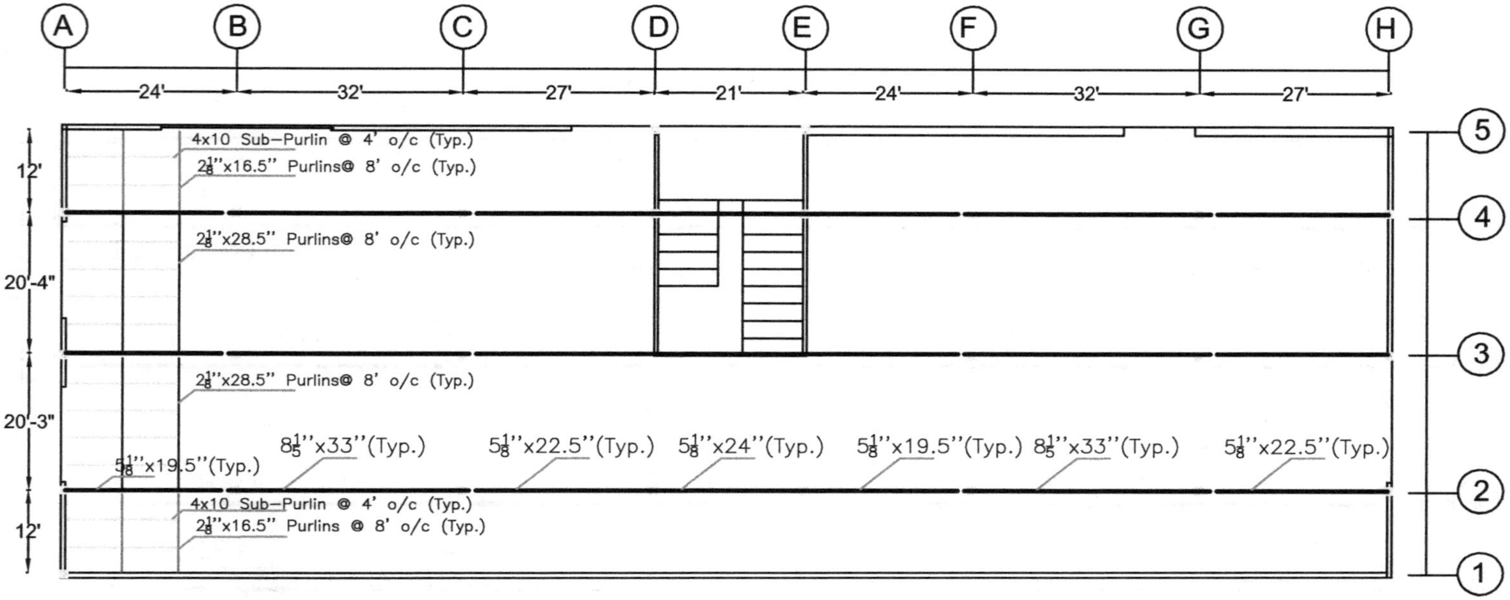
A
B
C
D
E
F
G
H
24'
32'
27'
21'
24'
32'
27'
1
2
3
4
5
12'
20'-4"
20'-3"
12'
4x10 Sub–Purlin @ 4' o/c (Typ.)
2 1/8"x16.5" Purlins@ 8' o/c (Typ.)
2 1/8"x28.5" Purlins@ 8' o/c (Typ.)
2 1/8"x28.5" Purlins@ 8' o/c (Typ.)
5 1/8"x19.5"(Typ.)
8 1/5"x33"(Typ.)
5 1/8"x22.5"(Typ.)
5 1/8"x24"(Typ.)
5 1/8"x19.5"(Typ.)
8 1/5"x33"(Typ.)
5 1/8"x22.5"(Typ.)
4x10 Sub–Purlin @ 4' o/c (Typ.)
2 1/8"x16.5" Purlins @ 8' o/c (Typ.)

Roof

REFERENCE

REFERENCE TITLE | REFERENCE No.

NOTES

Purlin and Sub–Purlin

Roof

KEY PLAN

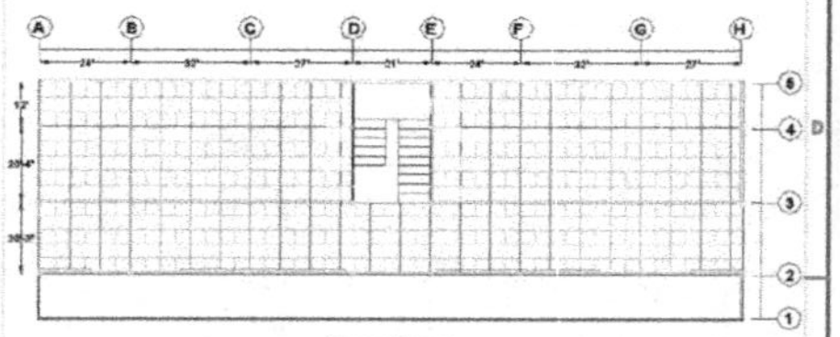

Second Floor

EQUIPMENT LIST

ABBREVIATIONS

REV.	DESCRIPTION	DRAWN	CHECKED	APPROVED	DATE
00	ISSUED FOR REVIEW/COMMENT	-	-	-	07 Jul 2018

CLIENT:

DESIGNER: ARZHANG ZAMANI

PROJECT: **Design of Wood Structures**

TITLE:

DOCUMENT NO.	CONTRACT NO.	DATE	ORIGINAL SIZE	SCALE	SHEET NO.	REV.
DOCUMENT	CONTRACT	DATE	A1	SCALE	X OF Y	REV.

66'

188'

6x10 DF-L (All Headers)

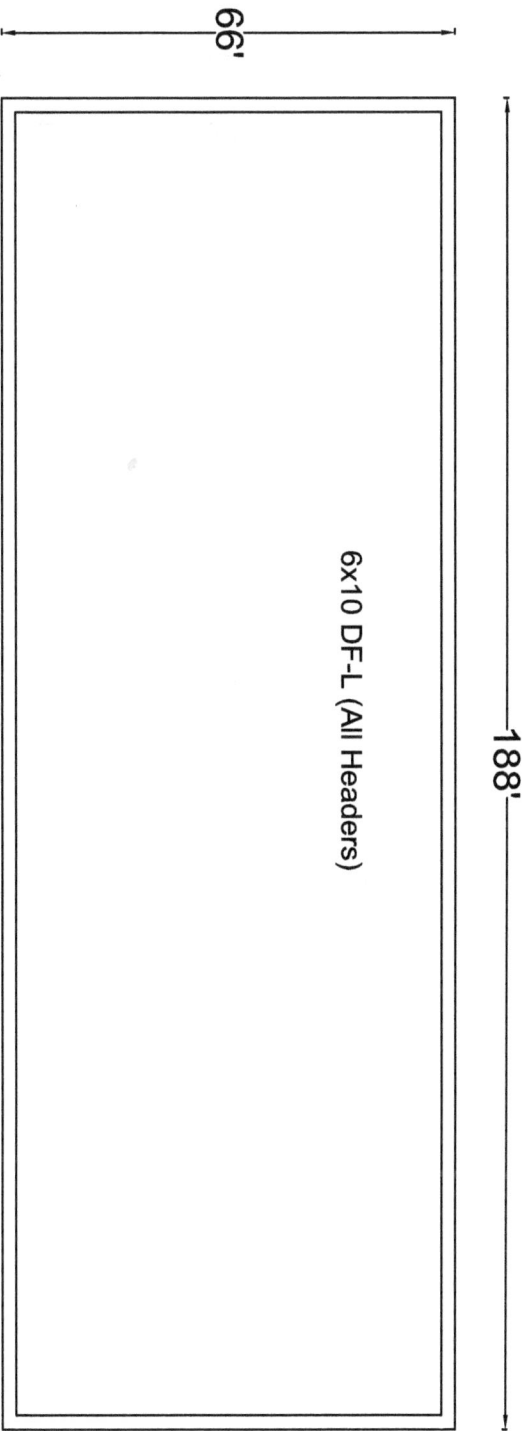

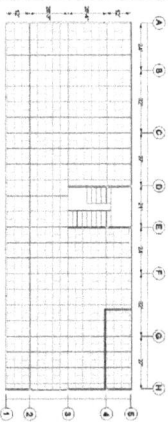

Design of Wood Structures

ARZHANG ZAMAN

Headers
Floor

KEY PLAN

REFERENCE TITLE

REFERENCE

REFERENCE No.

NOTES

ABBREVIATIONS

EQUIPMENT LIST

First Floor

DOCUMENT NO.   CONTRACT   DATE   SCALE   SHEET NO.   REV.

DRAWN   CHECKED   APPROVED

ISSUED FOR DEVELOPMENT   DESCRIPTION   REV.   CLIENT

PROJECT

Title

A1   X   OF   Y

07 Jul 2016

52'-6"

188'

6x12 DF-L (All Headers)

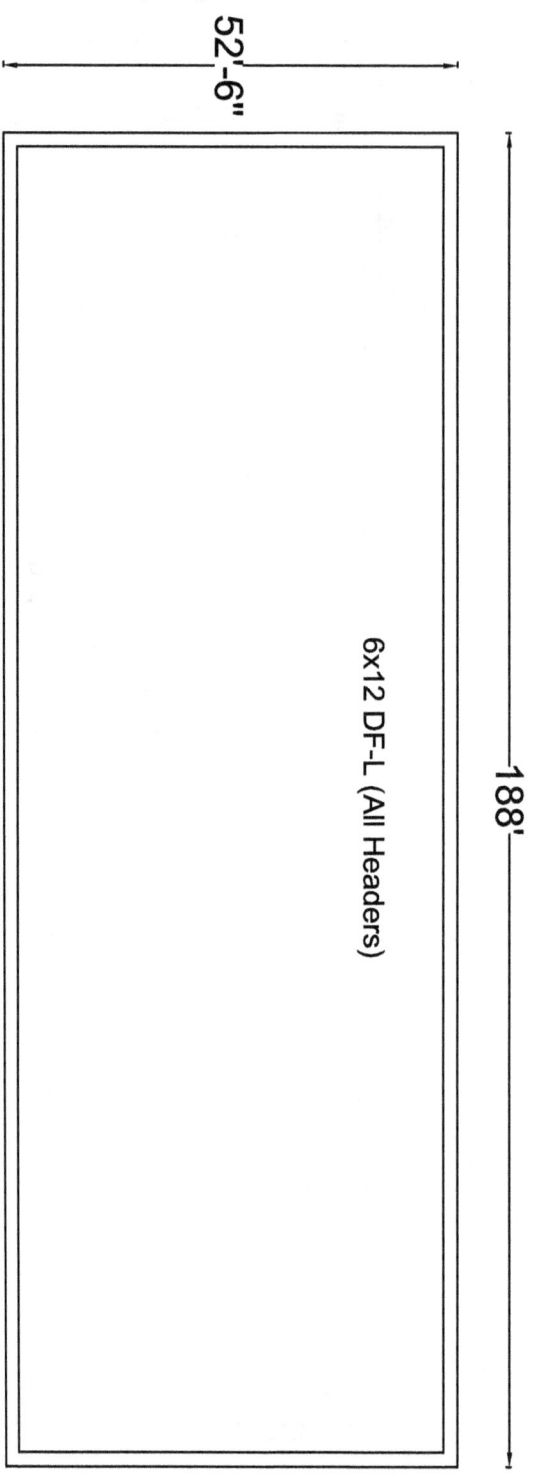

REFERENCE TITLE

REFERENCE

REFERENCE No.

NOTES

Headers

Roof

KEY PLAN

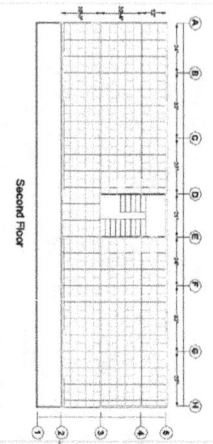

Second Floor

EQUIPMENT LIST

ABBREVIATIONS

Design of Wood Structures

DESCRIPTION

REFERENCE No.

DOCUMENT NO.   CONTRACT   DATE   DATE

DRAWN   APPROVED

DESIGNED   CHECKED

DATE   SCALE   X OF Y   REV.

First Floor

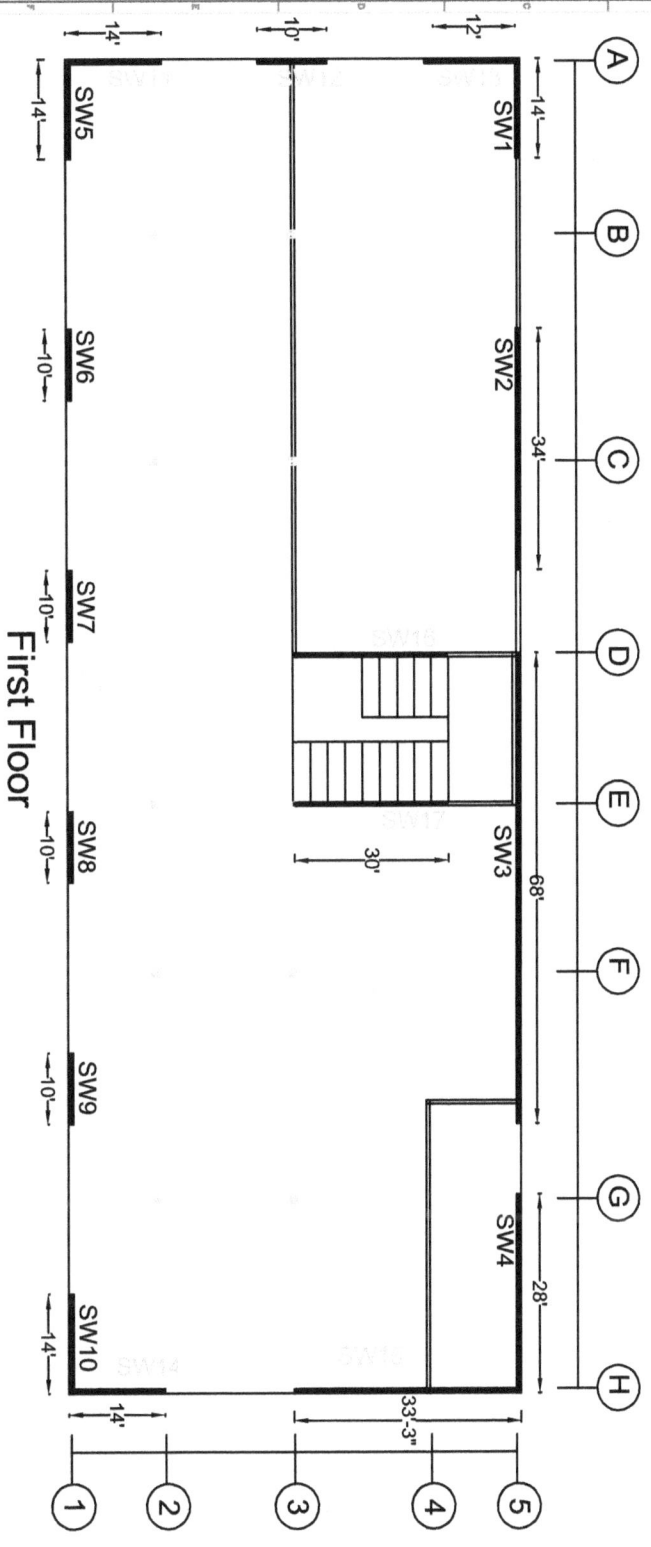

SW1   SW2   SW3   SW4

SW5   SW6   SW7   SW8   SW9   SW10

12'   10'   14'

14'   10'   10'   10'   10'   14'

14'

34'

68'

28'

30'

33'-3"

14'

A   B   C   D   E   F   G   H

1   2   3   4   5

Shear Walls
Floor

Design of Wood Structures

ARIHAND ZAMANI

REFERENCE, TITLE

REFERENCE

NOTES

REFERENCE, No.

KEY PLAN

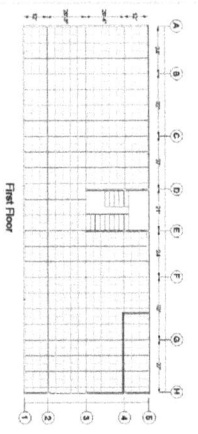

First Floor

EQUIPMENT LIST

ABBREVIATIONS

ISSUED FOR REVIEW/COMMENT

DOCUMENT NO.

CONTRACT

DATE

SCALE

SHEET NO.

REV.

Roof

SW9 SW10
10" 14'
14' SW1
14' SW5
SW2
34'
SW6 44'
SW12 23'
SW13
SW3 58'
SW7 44'
SW4 38'
SW8
SW11
33'-3"

A B C D E F G H
1 2 3 4 5

PRODUCED BY AUTODESK STUDENT VERSION

Design of Wood Structures

AIZWAD ZAMAN

Shear Walls
Roof

KEY PLAN

Second Floor

REFERENCE, TITLE   NOTES   REFERENCE   REFERENCE, No.

DRAWING No.   CURRENT No.   DATE
DOCUMENT   CONTRACT   DATE

DESCRIPTION   DRAWN   CHECKED   APPROVED
ORIGINAL SIZE   SCALE   SHEET No.   REV.
A1   X OF Y

ABBREVIATIONS

EQUIPMENT LIST

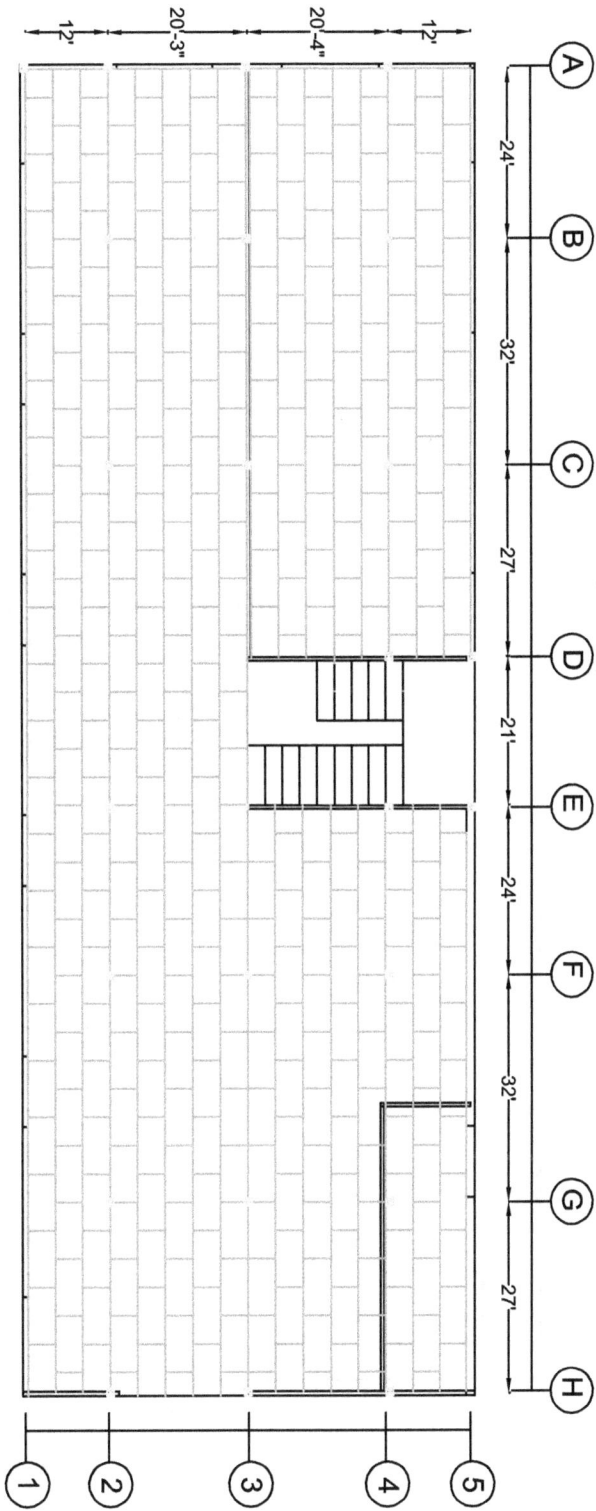

First Floor

Ply-Wood Layout

Floor

A B C D E F G H

24' 32' 27' 21' 24' 32' 27'

12' 20'-3" 20'-4" 12'

1 2 3 4 5

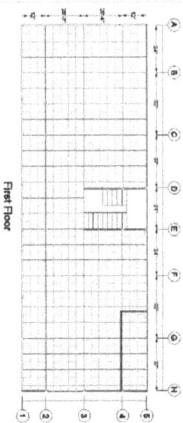

REFERENCE TITLE   REFERENCE

NOTES   REFERENCE No.

KEY PLAN

First Floor

EQUIPMENT LIST

EL.   ELEVATION

ABBREVIATIONS

ISSUED FOR DEVELOPMENT

DESCRIPTION

CLIENT

DOCUMENT No.   CONTRACT   DATE   ORIGINAL SIZE   SCALE   SHEET No.   REV.
                                        A1   16   X OF Y   REV

AZZHANG ZAMAN

Design of Wood Structures

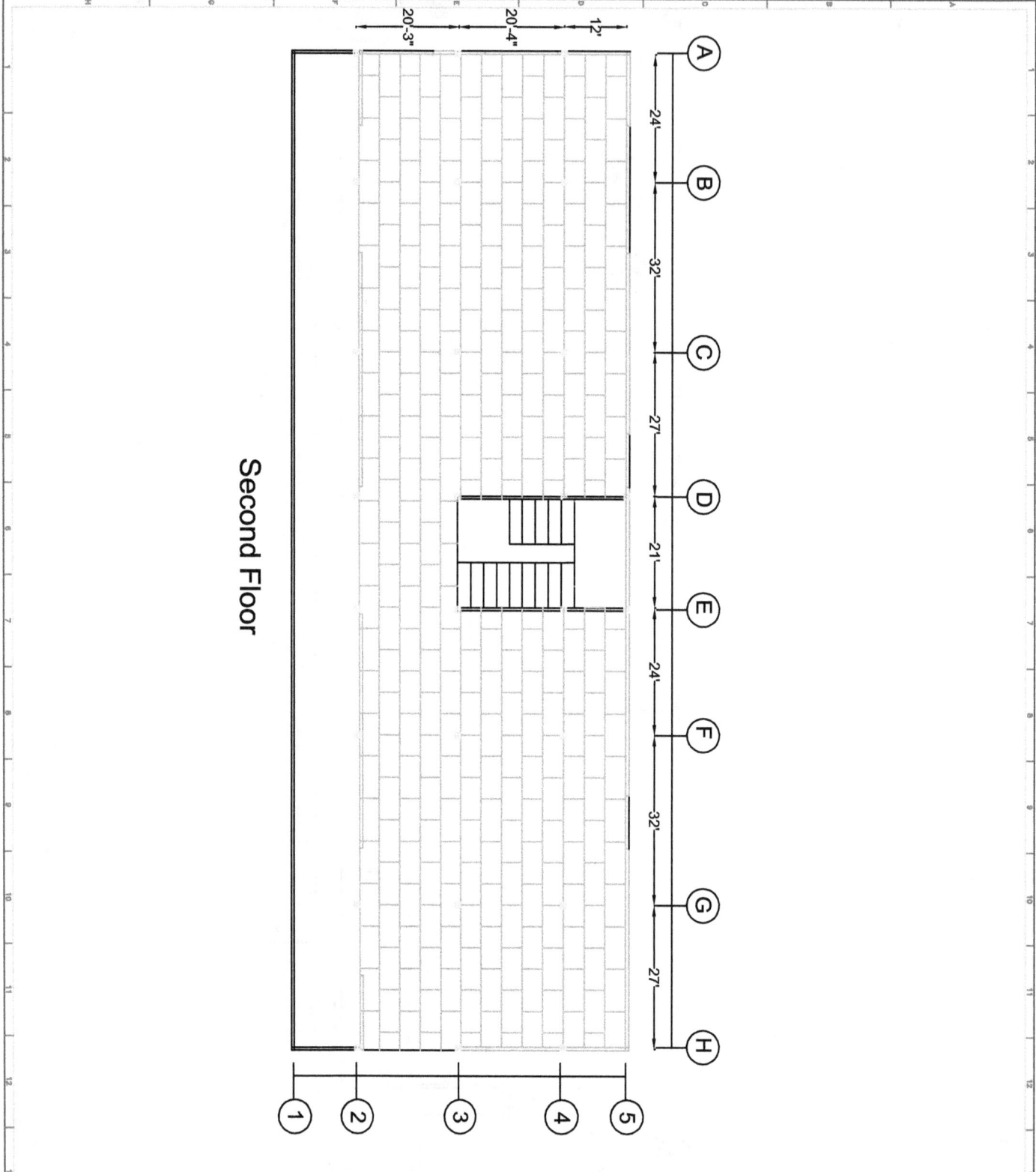

Second Floor

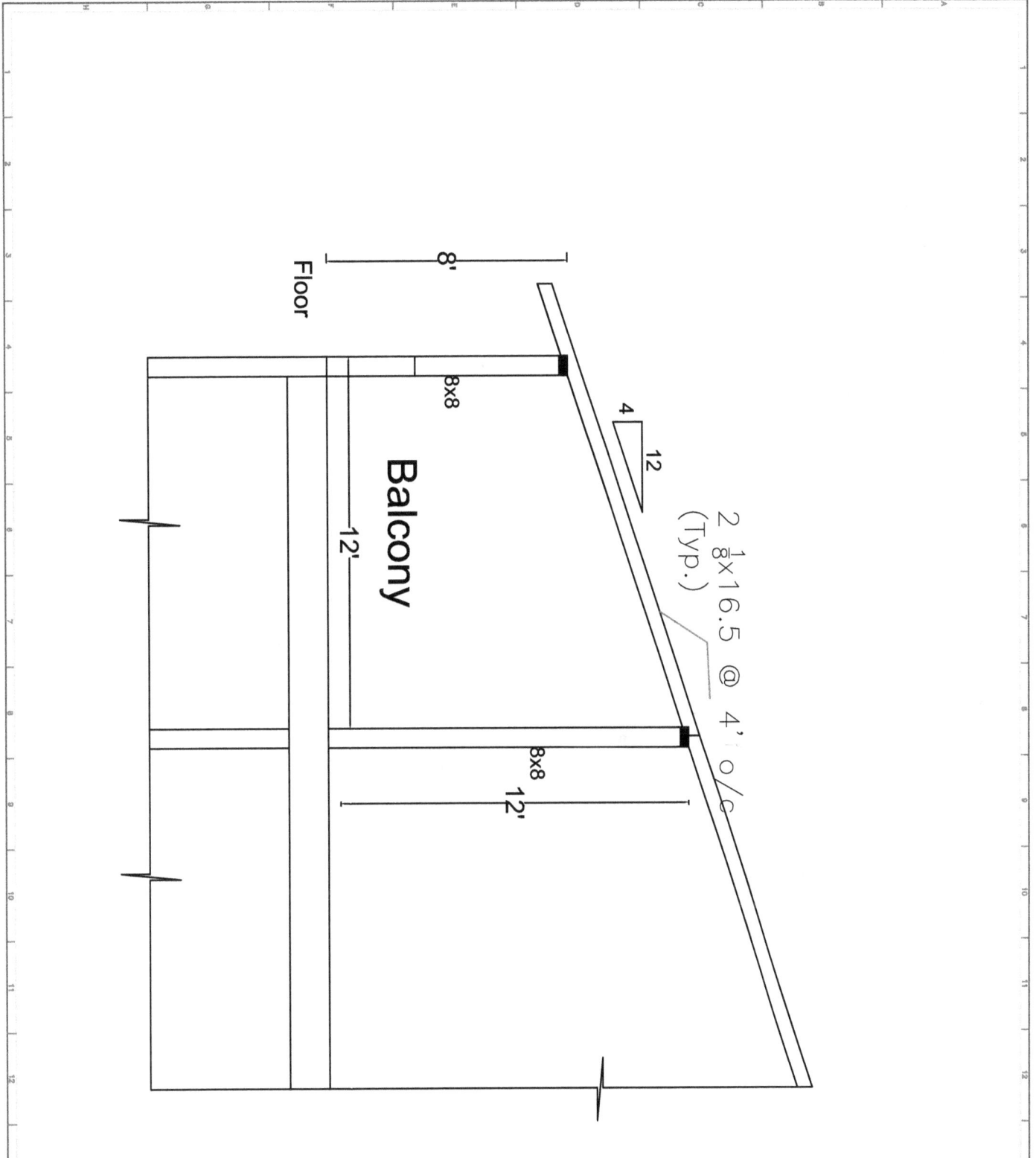
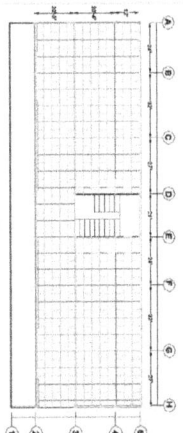

Loads:

In this project, building is analyzed and designed based on Dead Load, Live Load, Roof Live Load ,and Snow Load in addition to Lateral live load which is earthquake load.

Dead Loads:

Dead Load in the building consists of the weight of materials that are used for building. Based on floor detailing, and the intensity of load is calculated in Psf. The self-weight of members is added to the dead load. The density of wood is assumed to be 36 pcf for calculating the self-weight.

Roof:

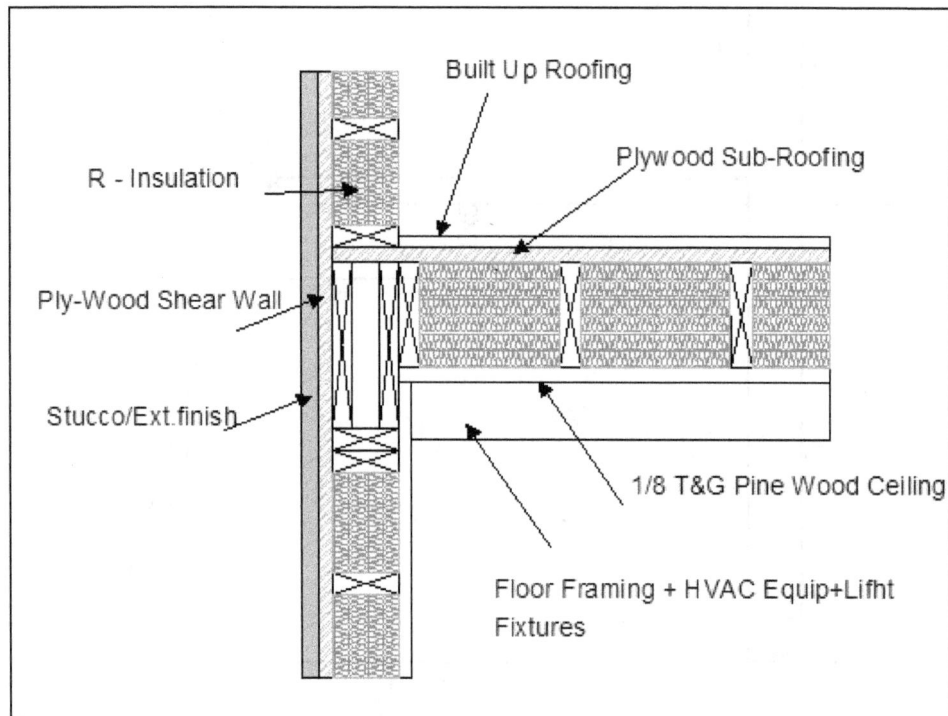

Above figure is typical detailing for the roof (Dead Load)

Roof

Dead (Type)	*Load (psf)*
"Built up flooring"	4
"Sub−flooring"	1.5
"R−19 Floor Insullation"	1.5
"1/8 T&G Pine Wood Ceiling"	4
"Misc + Light Fixtures"	5
"Total"	16

Therefore, the Base Load for Dead Load without considering the self-weight of member is 16 psf.

The self weight of members have to be added to above numbers. These nubers have been mentioned below based on designed values of fllowing sections.

Self.weight

$Sub-Purlin$ $1.889 \cdot psf$

$Purlin$ $3.968 \cdot psf$

Weight of GLBs in the floor:

$Total.Volume$ $78.81 \cdot ft^3 \cdot 3 = 236.43 \ ft^3$

$Weight$ $236.43 \ ft^3 \cdot 36 \ pcf = 8511.48 \ lbf$ $Area := 11708 \ ft^2$

GLB $\dfrac{8511.48 \ lbf}{Area} = 0.727 \ psf$

Therefore, Summary of loads:

$Roof$ $D_r := 16 \cdot psf + 1.889 \cdot psf + 3.968 \cdot psf + 0.727 \cdot psf = 22.584 \ psf$

Floor:

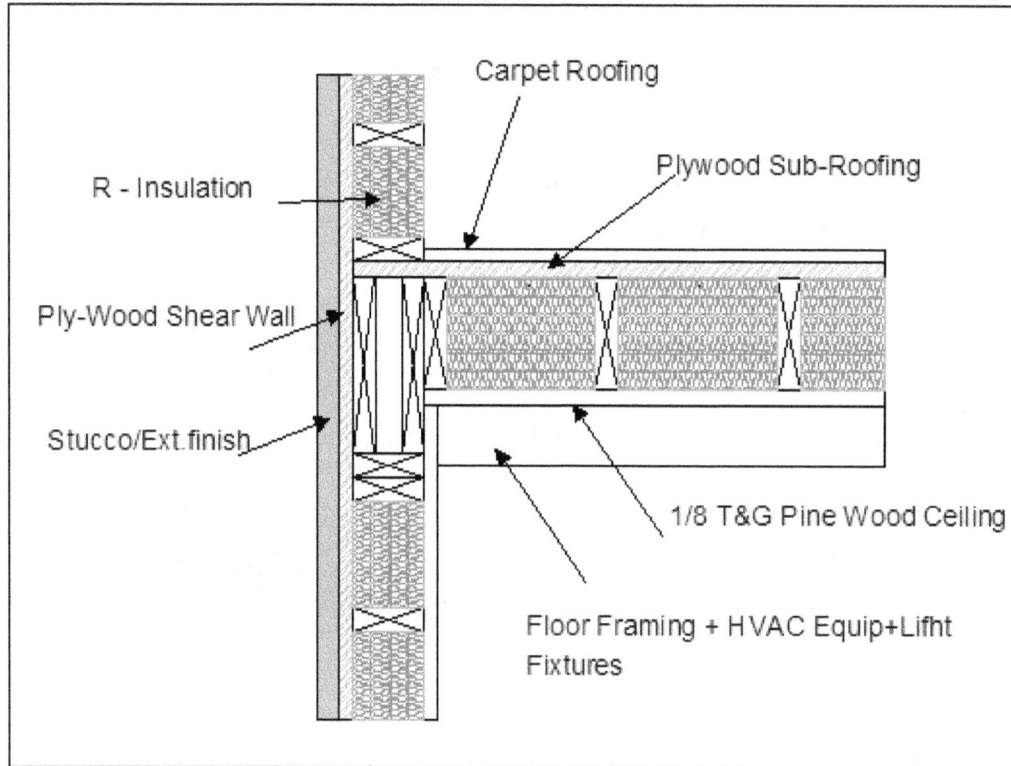

Dead (Type)	Load (psf)
"Carpet flooring"	4
"Sub−flooring"	1.5
"R−19 Floor Insullation"	1.5
"1/8 T&G Pine Wood Ceiling"	4
"Misc + Light Fixtures"	5
"Total"	16

Therefore, the Base Load for Dead Load without considering the self-weight of member is 16 psf.

The self-weight of members has to be added to above numbers. These numbers have been mentioned below based on designed values of following sections.

Self.weight

  Sub − Purlin $1.889 \cdot psf$

  Purlin $3.968 \cdot psf$

  Weight of GLBs in the floor:

    Total.Volume $78.81 \cdot ft^3 \cdot 3 = 236.43 \; ft^3$

    Weight $236.43 \; ft^3 \cdot 36 \; pcf = 8511.48 \; lbf$ $Area := 11708 \; ft^2$

      GLB $\dfrac{8511.48 \; lbf}{Area} = 0.727 \; psf$

  Therefore, Summary of loads:

    Floor $D_F := 16 \cdot psf + 1.889 \cdot psf + 3.968 \cdot psf + 0.727 \cdot psf = 22.584 \; psf$

Roof Live Loads:

Roof Live load is 16 psf . which is mainly for roof construction. The effect of roof slope is considered.

Snow Load:

Snow load is given to us for this particular building which is 65 psf. The effect of roof slope is considered.

Live Load:

Live Load is given as 50 psf.

Lateral Loads:

Earthquake load is used as a lateral load for designing of this building.
The Design equivalent earthquake load is given at each floor as following:

in   two   directions for   each   floor

Floor	Vf (kip)
	65.48

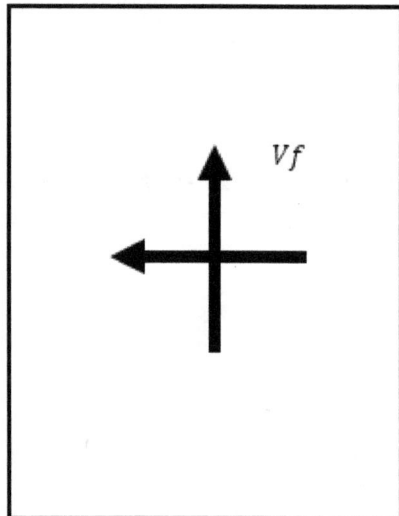

Floor

Shear $65.48 \cdot kip$

Roof	Vr (kip)
	116.93

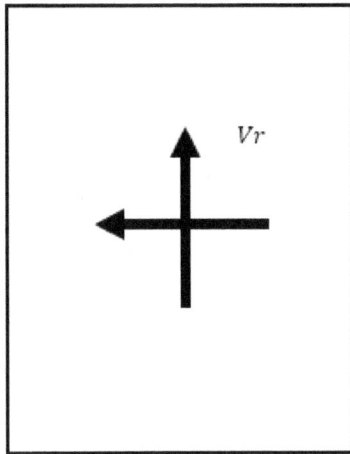

Roof

Shear $116.93 \cdot kip$

Building's view

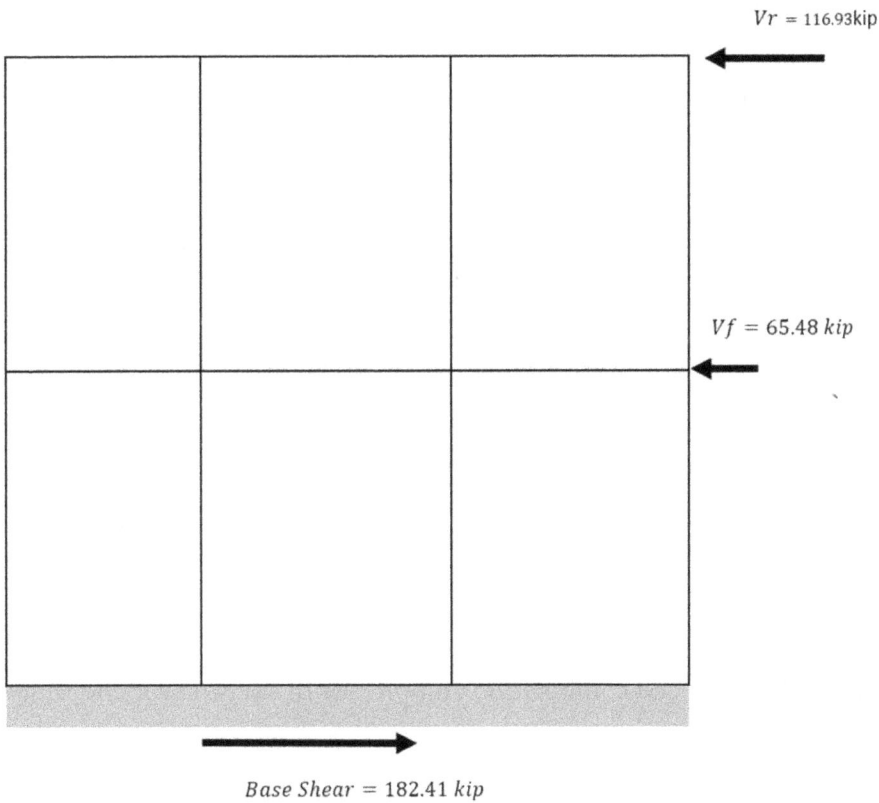

$Vr = 116.93 \text{kip}$

$Vf = 65.48 \ kip$

Base Shear $= 182.41 \ kip$

Applicable Load Combination:

ASD governing load combinations according to IBC 1605.3.1 are:

Load Combination
D
D+L
D+L+(LR or S or R)
D+(0.7E)+L+(LR or S or R)
0.6D+W
0.6D+0.7E

D : Dead Load
L: Live Load
S: Snow Load
W: Wind Load
E: Earthquake Load
LR: Roof Live Load
R: Rain Load

As it was mentioned the applicacle loads to this buidings are as below:

D
L
LR
S
E

In the design of each member, the load combination and process of decision for choosing the most governing case is delivered.

Summary of Glu Lam Beams Design: "FLOOR"

ALL   OF   GLBs   Floor

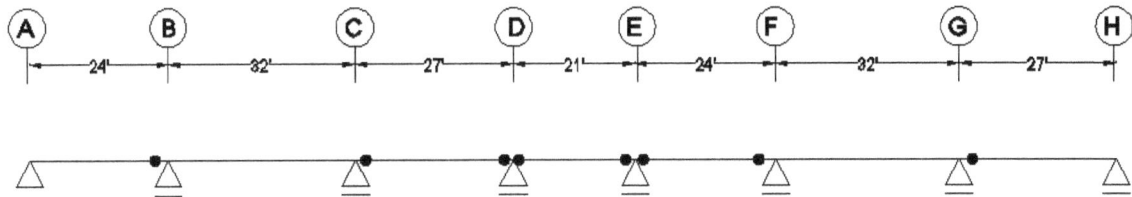

USE, GLBs as:

Left Cantilever "AB & EF " 51/8"x19.5"   (Length:16.8ft)

middle two cantilever " BC&FG " 8.5"x31.5" (Length:47.3ft)

Right Cantilever "DC&GH" 51/8"21" (Length:18.9ft)

simply supporte "DE" 51/8"x24" (Length:21ft)

Please check section C , for layout and GLBs' plan.

"Floor Glu Lam Beams"

GLBs along line 3 and 4 have the highest tributary area and demand. Therefore, the Glbs for first floor will be used from design of these lines. The GLBs along line 3 is specified in section of C this report. And it can also e found on the GLBs' layout plan.

Loads

$Dead\,(Type)$	$Load$
	(psf)
"Carpet flooring"	4
"Sub−flooring"	1.5
"R−19 Floor Insullation"	1.5
"1/8 T&G Pine Wood Ceiling"	4
"Misc + Light Fixtures"	5
"Total"	16

$Live$	$Load$
	(psf)
"Floor Live Load"	50

Governing.Load.Combination

$$D+L \qquad C_D = 1.0$$

According to the design of Purlin and sub-purlins their weights have to be added to design the GLBs.

Glu-Lam Beam (Load Calculation):

Floor

D1	$16 \cdot psf$
L	$50 \cdot psf$

Sub.Purlin

Size   4 $x \cdot 10$ 48 $''(o.c)$ $A_{s.p} := 3.5 \cdot in \cdot 9.25 \cdot in = 32.375 \ in^2$

$G := 0.5$ $mc := 15$ $S := 48 \cdot in$

$$\rho := 62.4 \cdot \left(\frac{G}{1 + G \cdot (0.009) \cdot mc} \right) \left(1 + \frac{mc}{100} \right) \cdot pcf = 33.611 \ pcf$$

$$W_{s.p} := \rho \cdot A_{s.p} \cdot \left(\frac{12 \cdot in}{S} \right) \cdot \frac{1}{ft} = 1.889 \ psf$$

Weight purlin in psf

Size $2 \, \frac{'1}{8} \, x \cdot 28.5$ 8 $'\,(o.c)$ $A_p := 2.5 \cdot in \cdot 28.5 \cdot in = 71.25 \ in^2$

$G := 0.5$ $mc := 15$ $S := 8 \cdot ft$

$$\rho := 62.4 \cdot \left(\frac{G}{1 + G \cdot (0.009) \cdot mc} \right) \left(1 + \frac{mc}{100} \right) \cdot pcf = 33.611 \ pcf$$

$$W_p := \rho \cdot A_p \cdot \left(\frac{12 \cdot in}{S} \right) \cdot \left(\frac{1}{ft} \right) = 2.079 \ psf \qquad\qquad W_p + W_{s.p} = 3.968 \ psf$$

Tributary Area:

Glulam.Beam $20\,' - 4\,''$ *along.line* (3) (4)

The biggest size member is chosen for weights.

Glu-Lam Beam Design:

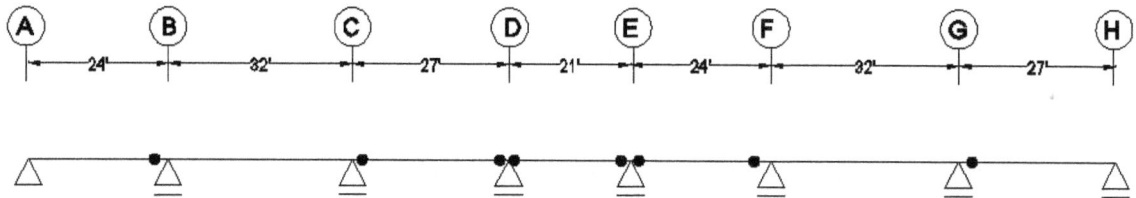

As can be noticed from the layout of Glu-lam beam, total system can be divided to two cantilevered system (ABCD & EFGH), and one simply supported DE.

Optimized alpha will be chosen to determine the cantilevered part and location of hinges and connections.

Design Process :
The most critical line of glu-Lam beam is designed for floor and roof.

Floor

        "AB & EF , BC&FG , DC&GH"   "DE"

   Floor

      D1 $16 \cdot psf + 3.968 \cdot psf = 19.968 \; psf$ $Width = 20.33 \;'$

      L $50 \cdot psf$

In order to get the most optimum result, different alphas have been tested to get the lowest volume of GLB.

	Cantilever ratio (a)	Total Volume(ft^3)	AB		BC		CD	
			b(in)	d(in)	b(in)	d(in)	b(in)	d(in)
1	0.195	82.22	5.13	22.5	8.5	31.50	5.13	25.50
2	0.26	80.01	5.125	21	8.5	31.5	5.125	22.5
3	0.24	80.76	5.125	21	8.5	31.5	5.125	24
4	0.2	81.14	5.125	22.5	8.5	31.5	5.125	24
5	0.15	82.70	5.125	24	8.5	31.5	5.125	25.5
(Governs)	0.3	78.81	5.125	19.5	8.5	31.5	5.125	21

for.the.floor

Therefore,
Left Cantilever "AB & EF " 51/8"x19.5" (Length:16.8ft)
middle two cantilever " BC&FG " 8.5"x31.5" (Length:47.3ft)
Right Cantilever "DC&GH" 51/8"21" (Length:18.9ft)
simply supporte "DE" 51/8"x24" (Length:21ft)

Deflection Limits							b	5.125	in
Ceiling type	Plaster	P					d	24.00	in
Deflection L	360						Sx	492.00	in
Deflection D+L	240						fb	1,954	psi
LOADING			weigth				check	OK	
wD	36.39	405.94944	30.75	plf					
wL	84.71	1016.5	plf						
wS	0.00	0.0008	plf						
Duration Factor	ASD (D)	ASD (D+L)	ASD (D+S)				Deflection Limits		
CD	0.9	1	1.25					Calculated	Limits Deflection
Govern ASD ?	485.22	1453.20	0.00	plf			L	0.63	0.70	in	OK
		Governs					D+L	0.82	1.05	in	OK
Beam Dimensions									
b	5.125	in	ro(pcf)	36.00					
Laminations t	1.5	in	A(in^2)	123.00					
n laminations	16		A(ft^2)	0.85					
d	24.00	in	W(glulam)(plf)	30.75					

Sample.of.calculations

Two.Cantilever.System   *Mid.span.Beam*

Load.Combinations

Deflection Limits			
Ceiling type	Plaster	P	
Deflection L	360		
Deflection D+L	240		

LOADING			weigth	
wD	39.41	405.94944	66.94	plf
wL	84.71	1016.5	plf	
wS	0.00	0.001	plf	

Duration Factor	ASD (D)	ASD (D+L)	ASD (D+S)	
CD	0.9	1	1.25	
Govern ASD ?	525.43	1489.39	378.31	plf
		Governs		

Beam Dimensions				
b	8.500	in		
Laminations t	1.5	in	ro(pcf)	36.00
n laminations	21		A(in^2)	267.75
d	31.50	1	A(ft^2)	1.86
			W(glulam)(plf)	66.94

Beam Length				
L	384.00	32	ft	Mid span
LA	288.00	24	ft	VOLUME(BC)ft^3
LB	324.00	27	ft	87.95
a1	0.300		L=	47.30

Adjustment.Factors

Adjustment Factors	Fb	Fv	F'c+	E
CM	1.00	1.00	1.00	1.00
Ct	1.00			
CL	0.99	Need to check for supports later		
CF (not for GLB)	1.00			
xyz	20			
k	1			
CV (only in GLB)	0.91	Note: Do not use CL and CV together		
Cfu	1.00			
Cr	1.00			
Cc	1.00			
Cf	1.00			

Bending & Deflection

Design For Bending		
Mmax	1954374	lbs-in
CD	1.00	
CM	1.00	
Ct	1.00	
CL	0.99	
CF	1.00	
CV	0.91	**Note:** Do not use CL and CV together
Cfu	1.00	
Cr	1.00	
Fb	2,400	psi
F'b	2,162	psi
b	8.500	in
d	31.50	in
Sx	1405.69	in
fb	1,390	psi
check	**OK**	

Deflection Limits				
	Calculated	Limits	Deflection	
L	0.93	1.07	in	**OK**
D+L	0.99	1.60	in	**OK**

Shear

Design For Shear		
Vmax	26139	lbs
w	124.12	lbs/in
b	8.500	in
d	31.50	in
Area	267.75	in^2
V'	22229	lbs
CD	1.00	
CM	1.00	
Ct	1.00	
CH	1.00	
Fv	300	
F'v	300	psi
fv	146	psi
check	**OK**	

L/d = 12

Bearing

	Design For Bearing		
R	51037	lbs	
CM	1		
Ct	1		
Cb	1	Need to check supports later	
Fc+	650	psi	
F'c+	650	psi	
b	8.500	in	CCO31/
lb	9.24	in	
Use Base PL	CCO5 1/4-6 ECC		
R cap	15785	lbs	
Check Plate	**NG**		
Use a lb	10	in	
fv	600	psi	
Cb	1.00		
F'c+	650	psi	
check	**OK**		

Bracing

Design for bracing

lu-top	48.00	4	ft	Assume purlines/joits at 4ft on center.
lu-bottom	96.00	8	ft	max of l1 and l2 times 2
b	8.500	in		
d	31.50	in		
Design for stability				
E'y	1,700,000	psi		
	GLB			
Kbe	0.610			
lu/d top	1.52			
lu/d botto	3.05			

$$l_e = \begin{cases} 2.06\,l_u & l_u/d < 7 \\ 1.63\,l_u + 3d & 7 \le l_u/d \le 14.3 \\ 1.84\,l_u & l_u/d > 14.3 \end{cases}$$

le top	98.88		
le bottom	197.76		
Rb top	6.57	**OK**	
Rb bottom	9.29	**OK**	

$$R_B = \sqrt{\frac{l_e d}{b^2}} \le 50$$

FbE-top	24055	psi
FbE-bottc	12027	psi
Fb*	2376	psi	Remove the CV since CL is not to be used with CL

Q-top	10.12	
Q-bottom	5.06	

$$F_{bE} = \frac{K_{bE} E'_y}{R_B^2}$$

CL top	0.99	
CL bot	0.99	

$$Q = \frac{F_{bE}}{Fb^*}$$

F'b top	2363	psi
F'b bot	2348	psi

$$C_L = \left[\left(\frac{1+Q}{1.9}\right) - \sqrt{\left(\frac{1+Q}{1.9}\right)^2 - \frac{Q}{0.95}}\right] \le 1.0$$

fb	1390	psi
OK		
OK		

Note: For all of the designed beams all requirements have been checked. GLB is laid out in so as to have the bending design control rather than deflection control. Above calculations have been repeated for all of the members.

Floor   ABCD

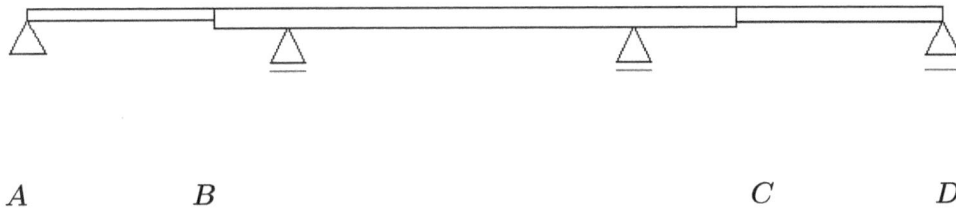

Here, portion of beam layout is shown.

Connections:

Connection of GLB to column caps.
ECCU

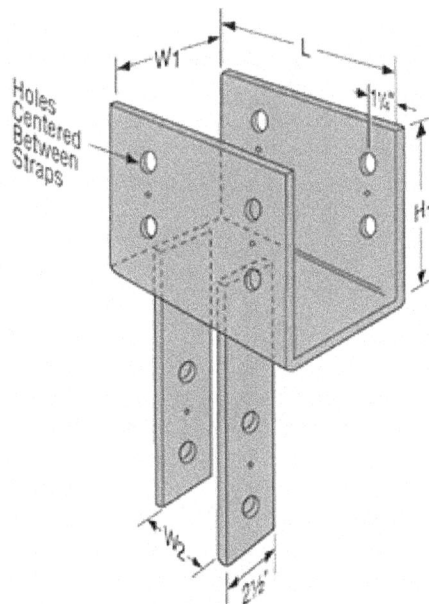

Different Connection

AB – 3   "GLB to column cap"

	Design For Bearing		
R	7804	lbs	
CM	1		
Ct	1		
Cb	1	Need to check supports later	
Fc+	650	psi	
F'c+	650	psi	
b	5.125	in	CCO31/
lb	2.34	in	
Use Base PL	CCO5 1/4-6 ECC		
R cap	15785	lbs	
Check Plate	**OK**		
Use a lb	2.5	in	
fv	609	psi	
Cb	1.15		
F'c+	748	psi	
check	**OK**		

USE   ECCU CC51/4 - 6

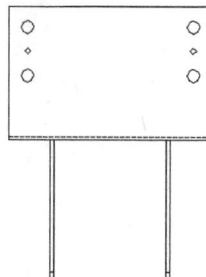

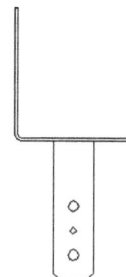

GLB.to.GLB

$DE-3$ *to* $EF-3$

                           USE   *HCCTE*

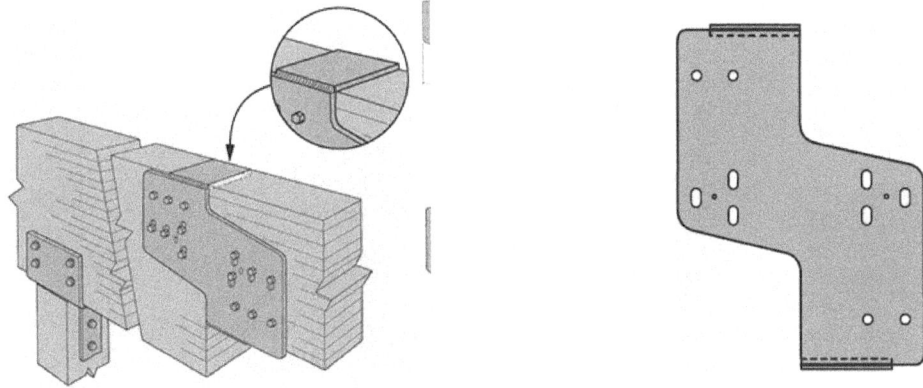

     "HCCTE5−7 H=21′′′" $l_b = 3 \ in$

Connection of purlins/joists to GLB:

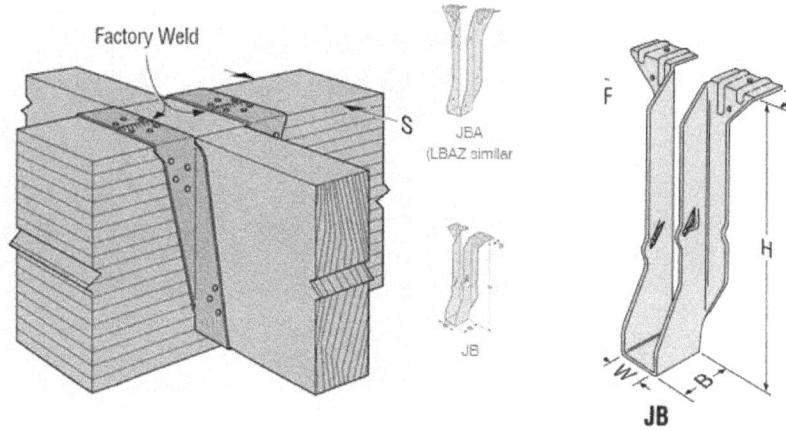

USE   *JB214A*   2 1/8"x13.5"

OR

USE

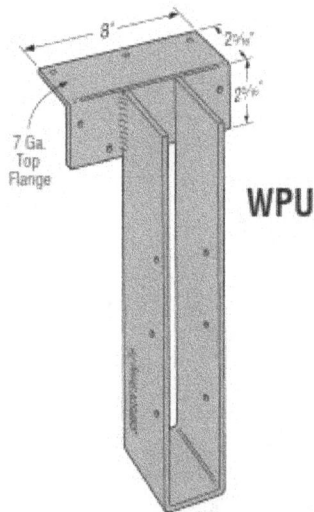

WPU   2 1/8"x13.5" - B=3.5"

Connection of Sub-Purlin to Purlin:

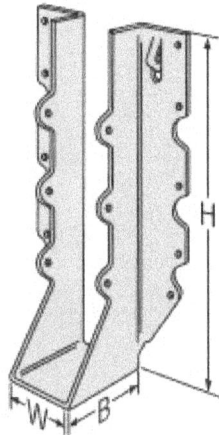

pg.83   of   simpson.catalog

"joist size"   4 *x*10

USE   4 *x*10   *U*410

or   use

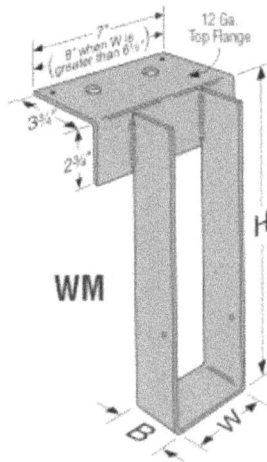

WM   4x10 - B=3.5"

Note: Simpson connections have been used for connection details. Please check the references for more details.

Summary of Glu Lam Beams Design: "Roof"

ALL   OF   GLBs   Roof

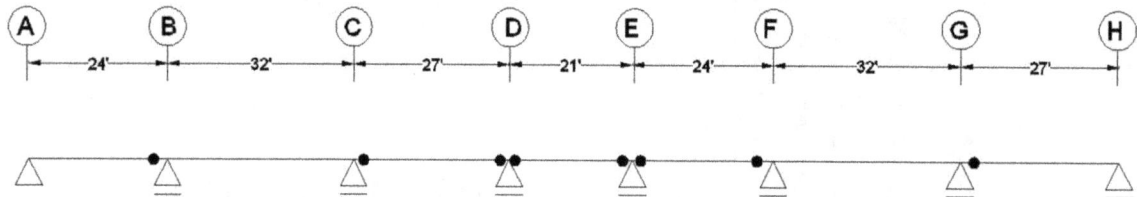

USE, GLBs as:

Left Cantilever "AB & EF " 51/8"x19.5"

middle two cantilever " BC&FG " 8.5"x33"

Right Cantilever "DC&GH" 51/8"x22.5"

simply supporte "DE" 51/8"x19.5"

Please check section C , for layout and GLBs' plan.

"Roof Glu Lam Beams"

GLBs along line 3 and 4 have the highest tributary area and demand. Therefore, the GLBs for Roof will be used from the design of these lines. The GLBs along line 3 is specified in the section of C this report. And it can also be found on the GLBs' layout plan.

Applicable.Loads

Dead $(Type)$	*Load* (psf)
"Built up flooring"	4
"Sub–flooring"	1.5
"R–19 Floor Insullation"	1.5
"1/8 T&G Pine Wood Ceiling"	4
"Misc + Light Fixtures"	5
"Total"	16

Roof.Live	*Load* (psf)
"Roof Live Load"	16

Snow	*Load* (psf)
"Snow Load"	65

Note: The load is converted to horizontal equivalent

Governing.Load.Combination

$$D+S \qquad C_D = 1.15$$

Glu-Lam Beam (Load Calculation):

Roof

    D1 $16 \cdot psf$

    L_r $16 \cdot psf$

    S $65 \cdot psf$

Sub.Purlin

    Size $4\ x \cdot 10\ \ 48\ ''(o.c)$ $A_{s.p} := 3.5 \cdot in \cdot 9.25 \cdot in = 32.375\ in^2$

    $G := 0.5$ $mc := 15$ $S := 48 \cdot in$

$$\rho := 62.4 \cdot \left(\frac{G}{1 + G \cdot (0.009) \cdot mc}\right)\left(1 + \frac{mc}{100}\right) \cdot pcf = 33.611\ pcf$$

$$W_{s.p} := \rho \cdot A_{s.p} \cdot \left(\frac{12 \cdot in}{S}\right) \cdot \frac{1}{ft} = 1.889\ psf$$

Weight purlin in psf

    Size $2\frac{'1}{8}\ x \cdot 28.5\ \ 8\ '(o.c)$ $A_p := 2.5 \cdot in \cdot 28.5 \cdot in = 71.25\ in^2$

    $G := 0.5$ $mc := 15$ $S := 8 \cdot ft$

$$\rho := 62.4 \cdot \left(\frac{G}{1 + G \cdot (0.009) \cdot mc}\right)\left(1 + \frac{mc}{100}\right) \cdot pcf = 33.611\ pcf$$

$$W_p := \rho \cdot A_p \cdot \left(\frac{12 \cdot in}{S}\right) \cdot \left(\frac{1}{ft}\right) = 2.079\ psf \qquad W_p + W_{s.p} = 3.968\ psf$$

Tributary Area:

    Glulam.Beam $20\ '-4\ ''$ *aling.Lines*   (3) (4)

The biggest size member is chosen for weights.

Glu-Lam Beam Design:

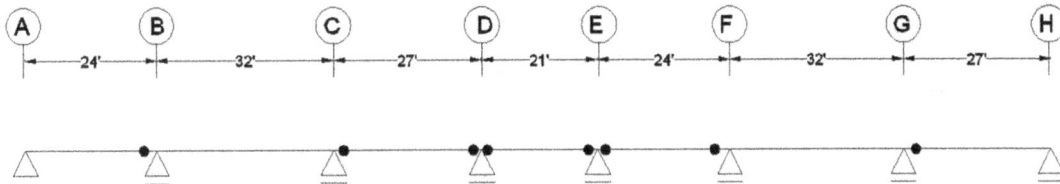

As can be noticed from the layout of Glu-lam beam, total system can be divided into two cantilevered system(ABCD & EFGH), and one simply supported DE

Optimized alpha will be chosen to determine the cantilevered part and location of hinges and connections.

Design Process :
The most critical line of glu-Lam beam is designed for floor and roof.

Roof

"AB & EF , BC&FG , DC&GH"   "DE"

Roof

D1 $16 \cdot psf + 3.968 \cdot psf = 19.968 \ psf$ $Width = 20.33 \ '$

L_r $16 \cdot psf$

S $65 \cdot psf$

Design Optimization by changing cantilever length:

	Cantilever ratio (a)	Total Volume(ft^3)	AB		BC		CD	
			b(in)	d(in)	b(in)	d(in)	b(in)	d(in)
1	0.2	80.26	5.125	18	8.5	34.5	5.125	21
2	0.22	79.68	5.125	18	8.5	34.5	5.125	19.5
3	0.24	80.25	5.125	18	8.5	34.5	5.125	19.5
4	0.26	79.87	5.125	16.5	8.5	34.5	5.125	19.5
5	0.28	81.90	5.125	16.5	8.5	36	5.125	18
6	0.3	82.58	5.125	16.5	8.5	36	5.125	18
7	0.18	80.77	5.125	19.5	8.5	34.5	5.125	21
8	0.16	78.11	5.125	19.5	8.5	33	5.125	21
(Governs)	0.14	77.65	5.125	19.5	8.5	33	5.125	21
10	0.12	78.46	5.125	19.5	8.5	33	5.125	22.5
11	0.15	79.10	5.125	19.5	8.5	33	5.125	22.5
12	0.145	79.00	5.125	19.5	8.5	33	5.125	22.5

for.the.roof

Therefore,
Left Cantilever "AB & EF " 51/8"x19.5"
middle two cantilever " BC&FG " 8.5"x33"
Right Cantilever "DC&GH" 51/8"x22.5"
simply supporte "DE" 51/8"x19.5"

Beam Type	24-V1 Sp/Sp				Fb	2,400	psi
MC limit	16	Note: 16 for GLB 19 Solid Sawn			F'b	2,694	psi
Moisture	16						
Density of GLB	36				b	5.125	in
Deflection Limits					d	19.50	in
Ceiling type	Plaster	P			Sx	324.80	in
Deflection L	360				fb	1,417	psi
Deflection D+L	240						
LOADING			weigth		check	**OK**	
wD	35.91	405.94944	24.98	plf			
wL	27.11	325.28	plf				
wS	0.00	0.0008	plf				

Deflection Limits				
	Calculated	Limits Deflection		
L	0.61	0.70	in	OK
D+L	0.96	1.05	in	OK

Duration Factor	ASD (D)	ASD (D+L)	ASD (D+S)
CD	0.9	1.25	1.15
Govern ASD ?	478.82	604.97	0.00 plf
		Governs	

Beam Dimensions					
b	5.125	in	ro(pcf)	36.00
Laminations t	1.5	in	A(in^2)	99.94
n laminations	13		A(ft^2)	0.69
d	19.50	in	W(glulam)(plf)	24.98

Beam Length

Design checks' sample:

Governing.Load

LOADING				weigth		
wD	39.67	405.94944		70.13	plf	
wL	27.11	325.28	plf			
wS	110.12	1321.45	plf			
Duration Factor	ASD (D)	ASD (D+L)	ASD (D+S)			
CD	0.9	1.25	1.15			
Govern ASD ?	528.97	641.08	1563.06	plf		
			Governs			
Beam Dimensions						
b	8.500	in				
Laminations t	1.5	in	ro(pcf)	36.00		
n laminations	22		A(in^2)	280.50		
d	33.00	in	A(ft^2)	1.95		
			W(ghulam)(plf)	70.13		
Beam Length						
L	384.00	32	ft	Mid span		

Adjustment.Factors

Adjustment Factors	Fb	Fv	F'c+	E
CM	1.00	1.00	1.00	1.00
Ct	1.00			
CL	1.00	Need to check for supports later		
CF (not for GLB)	1.00			
xyz	20			
k	1			
CV (only in GLB)	0.91	Note: Do not use CL and CV together		
Cfu	1.00			
Cr	1.00			
Cc	1.00			
Cf	1.00			

Bending

Design For Bending		
Mmax	3647879	lbs-in
CD	1.15	
CM	1.00	
Ct	1.00	
CL	1.00	
CF	1.00	
CV	0.91	**Note:** Do not use CL and CV together
Cfu	1.00	
Cr	1.00	
Fb	2,400	psi
F'b	2,505	psi
b	8.500	in
d	33.00	in
Sx	1542.75	in
fb	2,365	psi
check	**OK**	

Deflection

Deflection Limits			
	Calculated	Limits Deflection	
L	0.44	1.07	in	**OK**
D+L	0.60	1.60	in	**OK**

Shear

Design For Shear		
Vmax	21104	lbs
w	66.78	lbs/in
b	8.500	in
d	33.00	in
Area	280.50	in^2
V'	18900	lbs
CD	1.15	
CM	1.00	
Ct	1.00	
CH	1.00	
Fv	300	
F'v	345	psi
fv	113	psi
check	**OK**	

$$L/d = 12$$

Bearing

Design For Bearing		
R	57026	lbs
CM	1	
Ct	1	
Cb	1	Need to check supports later
Fc+	650	psi
F'c+	650	psi
b	8.500	in	CCO31 /
lb	10.32	in
Use Base PL	CCO5 1/4-6 ECC	
R cap	15785	lbs
Check Plate	**NG**	
Use a lb	6.75	in
fv	994	psi
Cb	1.00	
F'c+	650	psi
check	**NG**	

Bracing.C_L

Design for bracing				
lu-top	48.00	4	ft	Assume purlines/joits at 4ft on center.
lu-bottom	96.00	8	ft	max of l1 and l2 times 2
b	8.500	in		
d	33.00	in		
Design for stability				
E'y	1,700,000	psi		
	GLB			
Kbe	0.610			
lu/d top	1.45			
lu/d botto	2.91			
le top	98.88			
le bottom	197.76			
Rb top	6.72	OK		
Rb botton	9.50	OK		
FbE-top	22961	psi		
FbE-bottc	11481	psi		
Fb*	2760	psi		Remove the CV since CL is not to be used with CL
Q-top	8.32			
Q-bottom	4.16			
CL top	0.99			
CL bot	0.98			
F'b top	2741	psi		
F'b bot	2718	psi		
fb	2365	psi		
	OK			
	OK			

$$l_e = \begin{cases} 2.06\, l_u & l_u/d < 7 \\ 1.63\, l_u + 3d & 7 \le l_u/d \le 14.3 \\ 1.84\, l_u & l_u/d > 14.3 \end{cases}$$

$$R_B = \sqrt{\frac{l_e d}{b^2}} \le 50$$

$$F_{bE} = \frac{K_{bE} E'_y}{R_B^2}$$

$$Q = \frac{F_{bE}}{F b^*}$$

$$C_L = \left[\left(\frac{1+Q}{1.9}\right) - \sqrt{\left(\frac{1+Q}{1.9}\right)^2 - \frac{Q}{0.95}}\right] \le 1.0$$

Note: For all of the designed beam all requirements have been checked. GLB is laid out to have the bending design control rather than deflection control.

Roof   ABCD

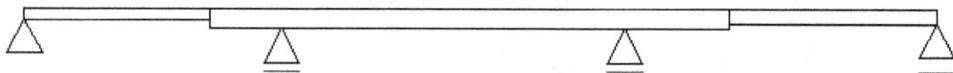

51/8"x16.5"   8.5"x33   51/8"x22.5" + DE+EFGH

Please, check section C for plan of GLBs.

Summary of DESIGN:

Use Braced 10x10 DF-L Select Structural (Floor Columns)

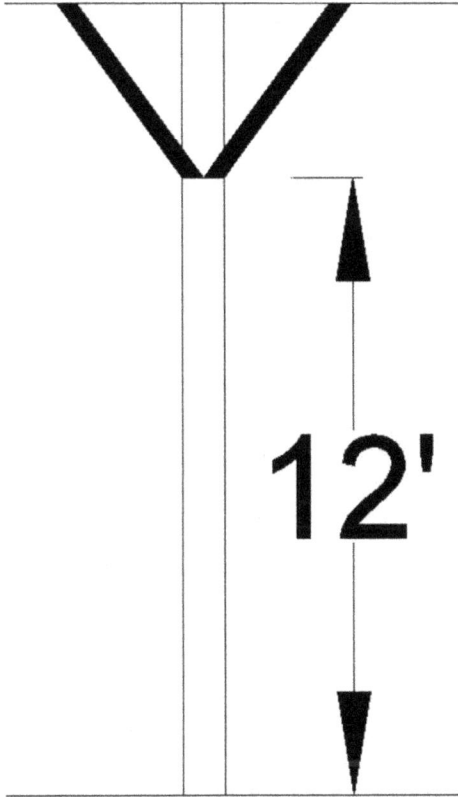

Floor   "Column C3"

Applicable loads:

P $(from.roof)$ $P := 52534.99 \cdot lbf$

Dead (Type)	Load (psf)
"Carpet flooring"	4
"Sub−flooring"	1.5
"R−19 Floor Insullation"	1.5
"1/8 T&G Pine Wood Ceiling"	4
"Misc + Light Fixtures"	5
"Total"	16

Live	Load (psf)
"Floor Live Load"	50

Self.weight

   $Sub-Purlin$ $1.889 \cdot psf$

   $Purlin$ $3.968 \cdot psf$

Weight of GLBs in the floor:

   $Total.Volume$ $78.81 \cdot ft^3 \cdot 3 = 236.43 \; ft^3$

   $Weight$ $236.43 \; ft^3 \cdot 36 \; pcf = 8511.48 \; lbf$ $Area := 11708 \; ft^2$

     GLB $\dfrac{8511.48 \; lbf}{Area} = 0.727 \; psf$

Therefore, Summary of loads:

Floor $D_r := 16 \cdot psf + 1.889 \cdot psf + 3.968 \cdot psf + 0.727 \cdot psf = 22.584 \ psf$

ASD   Load   Combination

LOADING	(PSF)		Area(ft^2)	Load(lb)	
wD	22.58		599.82	13546	
wL	50			29991	
wS	0.01			6	
Duration Factor	ASD (D)		ASD (D+L)	ASD (D+S)	
CD	0.9		1	1.15	
Govern ASD ?	15,051.48		29,991.00	5	lb
			Governs		

$D + L$ *Governs*

$$C_D := 1$$

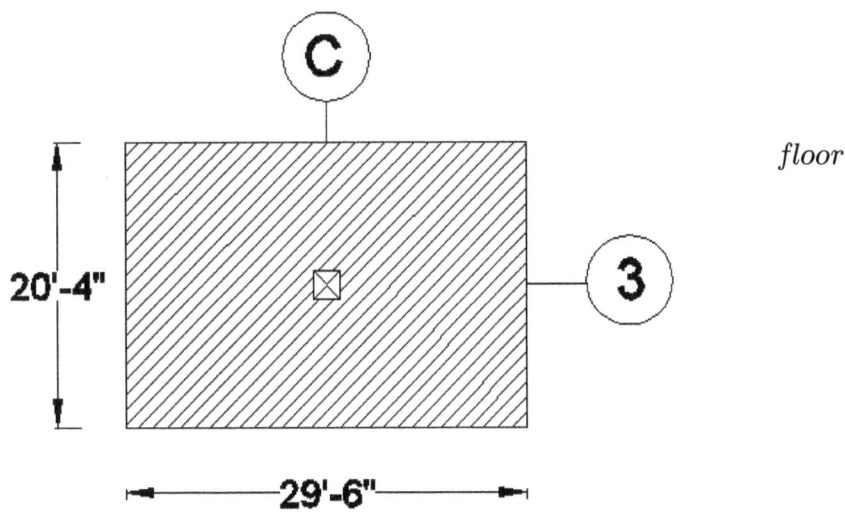

floor

$Tributary.Area := 29.5 \cdot ft \cdot (20.333) \cdot ft = 599.824 \ ft^2$

$D1 := 13546.425 \cdot lbf$

$L := 29991.2 \cdot lbf$

$P_1 := D1 + L = 43537.625 \ lbf$ NOTE:Load is sum of upper floor plus first floor.

$P_{floor} := P + P_1 = 96072.615 \ lbf$ $Design.Load$

$TRY.1$

USE $10 \ '' \ x10''$ $d := 9.5 \cdot in$ $L_u := 12 \cdot ft$

$F_c := 1300 \cdot psi$

$E_{min} := 620000 \cdot psi$ $E := 1700000 \cdot psi$

$C.Factors$

$C_M := 1$ $C_D = 1$

$C_f := \left(\dfrac{12 \cdot in}{d} \right)^{\left(\frac{1}{9} \right)} = 1.026$ $k := 1$ $k \cdot L_u = 12 \ ft$ $less.than.50$ ok

$C_f := 1$

$F_{c.star} := F_c \cdot C_D \cdot C_M \cdot C_f = 1300 \ psi$

$Solid.Sawn$ $c := 0.8$

$F_{CE} := \dfrac{0.822 \cdot E_{min}}{\left(\dfrac{k \cdot L_u}{d} \right)^2} = 2218.124 \ psi$

$C_P := \dfrac{1 + \dfrac{F_{CE}}{F_{c.star}}}{2 \cdot c} - \sqrt{\left(\dfrac{1 + \dfrac{F_{CE}}{F_{c.star}}}{2 \cdot c} \right)^2 - \dfrac{\dfrac{F_{CE}}{F_{c.star}}}{c}} = 0.838$

$$F'_c := F_{c.star} \cdot C_P = 1089.598 \ \textbf{\textit{psi}}$$

$$A_{req} := \frac{P_{floor}}{F_{c.star} \cdot C_P} = 88.173 \ \textbf{\textit{in}}^2$$

$$A_{available} := (9.5 \cdot \textbf{\textit{in}})^2 = 90.25 \ \textbf{\textit{in}}^2 \qquad ok \qquad \frac{A_{req}}{A_{available}} = 0.977 \qquad ok$$

Use Braced 10x10 DF-L Select Structural (Floor Columns)

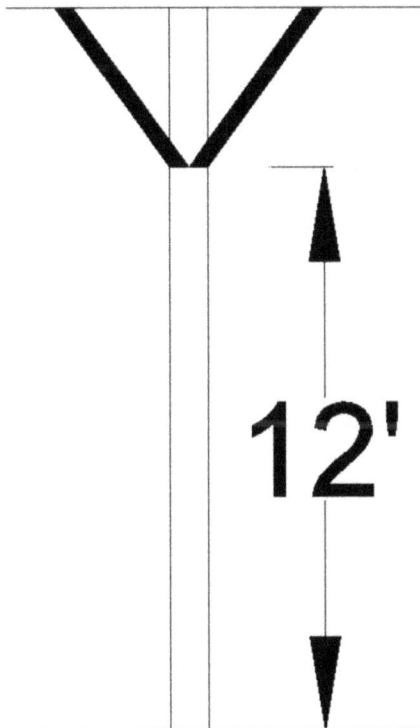

floor $10'' \, x10''$

Summary of DESIGN:

Use Braced 8x8 DF-L Select Structural (Roof Columns)

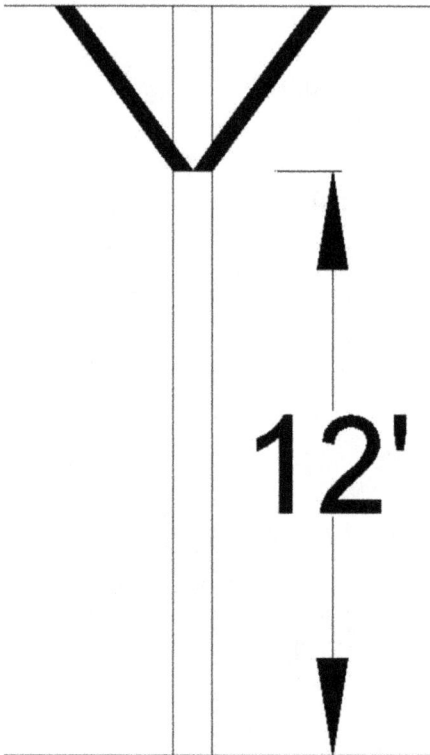

12'

Design of a Column:

Design Process:

For each floor two typical columns design are performed.One type column can be chosen from perimeter that is going to be a typical for all of perimeter columns in that floor and the other is going to be the one with the highest tributary area.
Since, the column is one of the most fundumental members in building.It can be used as a typical for all floor.

ROOF   "Column C3"

Applicable loads:

Dead (Type)	*Load (psf)*
"Built up flooring"	4
"Sub−flooring"	1.5
"R−19 Floor Insullation"	1.5
"1/8 T&G Pine Wood Ceiling"	4
"Misc + Light Fixtures"	5
"Total"	16

Roof.Live	*Load (psf)*
"Roof Live Load"	16

Snow	*Load (psf)*
"Snow Load"	65

Self.weight

Sub − Purlin	$1.889 \cdot psf$
Purlin	$3.968 \cdot psf$

Weight of GLBs in the floor:

$Total.Volume \quad 78.81 \cdot ft^3 \cdot 3 = 236.43 \ ft^3$

$Weight \quad 236.43 \ ft^3 \cdot 36 \ pcf = 8511.48 \ lbf \quad\quad Area := 11708 \ ft^2$

$GLB \quad \dfrac{8511.48 \ lbf}{Area} = 0.727 \ psf$

Therefore, Summary of loads:

$Roof \quad D_r := 16 \cdot psf + 1.889 \cdot psf + 3.968 \cdot psf + 0.727 \cdot psf = 22.584 \ psf$

$L_r := 16 \cdot psf$

$S := 65 \cdot psf$

ASD   Load   Combination

	LOADING	(PSF)		Area(ft^2)	Load(lb)	
	wD	22.58		599.8	13546	
	wL	16			9597	
	wS	65			38987	
	Duration Factor	ASD (D)		ASD (D+L)	ASD (D+S)	
	CD	0.9		1.25	1.15	
	Govern ASD ?	15,050.98		7,677.44	33,902	lb
					Governs	

D+S   Governs

$C_D := 1.15$

"Column with highest tributary area"

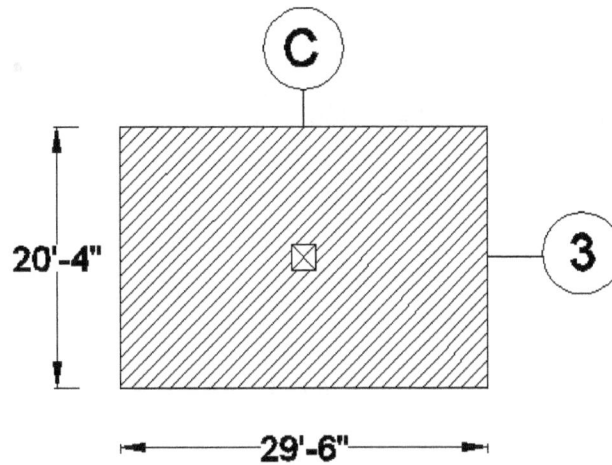

$$Tributary.Area := 29.5 \cdot ft \cdot (20.333) \cdot ft = 599.824 \; ft^2$$

$$D1 := 13546.425 \cdot lbf$$

$$S := 38988.56 \cdot lbf$$

$$P := D1 + S = 52534.985 \; lbf \quad Design.Load$$

Material used for column:
Square section
DF-L Select Structural
Ke=1 (V braced on top)
MC<15%

Column.Height

Column height from the v bracing to the top of GLB.

TRY.1

   USE   12 " x12" $d := 11.5 \cdot in$ $L_u := 12 \cdot ft$

$F_c := 1300 \cdot psi$

$E_{min} := 620000 \cdot psi$ $E := 1700000 \cdot psi$

C.Factors

   $C_M := 1$ $C_D = 1.15$

   $C_f := \left(\dfrac{12 \cdot in}{d} \right)^{\left(\frac{1}{9} \right)} = 1.005$ $k := 1$ $k \cdot L_u = 12 \; ft$ *less.than.*50   *ok*

   $F_{c.star} := F_c \cdot C_D \cdot C_M \cdot C_f = 1502.086 \; psi$

   Solid.Sawn $c := 0.8$

   $F_{CE} := \dfrac{0.822 \cdot E_{min}}{\left(\dfrac{k \cdot L_u}{d} \right)^2} = 3250.38 \; psi$

   $C_P := \dfrac{1 + \dfrac{F_{CE}}{F_{c.star}}}{2 \cdot c} - \sqrt{ \left(\dfrac{1 + \dfrac{F_{CE}}{F_{c.star}}}{2 \cdot c} \right)^2 - \dfrac{\dfrac{F_{CE}}{F_{c.star}}}{c} } = 0.88$

$F'c := F_{c.star} \cdot C_P = 1321.143 \; psi$

$A_{req} := \dfrac{P}{F_{c.star} \cdot C_P} = 39.765 \; in^2$

Let's try smaller member, for instance 8x8 .

TRY.2

    USE   8 ″ x8″ $d := 7.5 \cdot in$ $L_u := 12 \cdot ft$

    $F_c := 1300 \cdot psi$

    $E_{min} := 620000 \cdot psi$ $E := 1700000 \cdot psi$

    C.Factors

      $C_M := 1$ $C_D = 1.15$

      $C_f := \left(\dfrac{12 \cdot in}{d} \right)^{\left(\frac{1}{9} \right)} = 1.054$ *N.G* $k := 1$ $k \cdot L_u = 12 \; ft$ *less.than.50*

        $C_f := 1$

    $F_{c.star} := F_c \cdot C_D \cdot C_M \cdot C_f = 1495 \; psi$

      Solid.Sawn $c := 0.8$

    $F_{CE} := \dfrac{0.822 \cdot E_{min}}{\left(\dfrac{k \cdot L_u}{d} \right)^2} = 1382.487 \; psi$ $\dfrac{F_{CE}}{F_{c.star}} = 0.925$

$$C_P := \dfrac{1 + \dfrac{F_{CE}}{F_{c.star}}}{2 \cdot c} - \sqrt{\left(\dfrac{1 + \dfrac{F_{CE}}{F_{c.star}}}{2 \cdot c} \right)^2 - \dfrac{\dfrac{F_{CE}}{F_{c.star}}}{c}} = 0.663$$

    $F'_c := F_{c.star} \cdot C_P = 991.692 \; psi$

  $A_{req} := \dfrac{P}{F_{c.star} \cdot C_P} = 52.975 \; in^2$ $P = 52534.985 \; lbf$

    $A_{available} := \left(7.5 \cdot in \right)^2 = 56.25 \; in^2$ *ok* $\dfrac{A_{req}}{A_{available}} = 0.942$ *ok*

    Use Braced 8x8 DF-L Select Structural (ROOF Columns)

Summary of Purlin Design: "FLOOR"

USE, Purlins as below:

Type.1 $L = 20\,' - 4\,''$ *USE*   "21/8" x 22.5"   *GLB*   *DF − L* $MC = 15$

Type.2 $L = 12\,'$ *USE*   "21/8" x 13.5"   *GLB*   *DF − L* $MC = 15$

"Floor Purlin"

Note : Sub Purlin and Purlin have simply supported condition .

2 1/8" glu-lam joist 24F-V4 SP/SP - 36 pcf - M.C=15%

$$D \qquad 16 \cdot psf \cdot 8 \cdot ft = 128 \; plf$$

"self.weight:To Be Calculated"

$$L \qquad 50 \cdot psf \cdot 8 \cdot ft = 400 \; plf$$

$Type.1$　　　　　$Length$　　$20\,'-4\,''$

$Type.2$　　　　　$Length$　　$12\,'$

Note: The Purlins are categorized into two types that will fit two different lengths.
The design of beams is performed with the spreadsheet. Bending, shear, bracing, and deflection are checked.

$Type.1$　　　　　$Length$　　$20\,'-4\,''$

$A3-A2$　　　$(PLAN)$　　　$Check.Section.C$

$Type.2$　　　　　$Length$　　$12\,'$

$A5-A4$　　　$(PLAN)$　　　$Check.Section.C$

Design   of   Type.1   Length   20′−4″

Governing.Load.Combination

$D+L$ $C_D = 1.0$

LOADING			weigth	
wD	11.66	128	11.95	plf
wL	33.33	400	plf	
wS	0.00	0.008	plf	

Duration Factor	ASD (D)	ASD (D+L)	ASD (D+S)	
CD	0.9	1	1.15	
Govern ASD ?	155.50	539.95	121.71	plf
		Governs		

Beam Dimensions				
b	2.125	in		
Laminations t	1.5	in	ro(pcf)	36.00
n laminations	15		A(in^2)	47.81
d	22.50	in	A(ft^2)	0.33
			W(glulam)(plf)	11.95

Beam Length				
L	244.00	20.333	ft	Mid span
LA	0.00	0	ft	VOLUME(BC)ft^3
LB	0.00	0	ft	6.75

C.Factors

Adjustment Factors	Fb	Fv	F'c+	E
CM	1.00	1.00	1.00	1.00
Ct	1.00			
CL	1.00	Need to check for supports later		
CF (not for GLB)	1.00			
xyz	20			
k	1			
CV (only in GLB)	1.00	Note: Do not use CL and CV together		
Cfu	1.00			
Cr	1.00			
Cc	1.00			
Cf	1.00			

Bending

Design For Bending		
Mmax	334850	lbs-in
CD	1.00	
CM	1.00	
Ct	1.00	
CL	1.00	
CF	1.00	
CV	1.00	**Note:** Do not use CL and CV together
Cfu	1.00	
Cr	1.00	
Fb	2,400	psi
F'b	2,400	psi
b	2.125	in
d	22.50	in
Sx	179.30	in
fb	1,868	psi
check	**OK**	

Deflection

Deflection Limits				
	Calculated	Limits Deflection		
L	0.61	0.68	in	**OK**
D+L	0.76	1.02	in	**OK**

Shear

Design For Shear		
Vmax	5489	lbs
w	45.00	lbs/in
b	2.125	in
d	22.50	in
Area	47.81	in^2
V'	4477	lbs
CD	1.00	
CM	1.00	
Ct	1.00	
CH	1.00	
Fv	300	
F'v	300	psi
fv	172	psi
check	**OK**	

L/d = 11

Beaaring

Design For Bearing		
R	5489	lbs
CM	1	
Ct	1	
Cb	1	Need to check supports later
Fc+	650	psi
F'c+	650	psi
b	2.125	in
lb	3.97	in
Use Base PL	CCO5 1/4-6 ECC	
R cap	15785	lbs
Check Plate	**OK**	
Use a lb	6.75	in
fv	383	psi
Cb	1.00	
F'c+	650	psi
check	**OK**	

CCO31/

THEREFORE,

USE   "21/8" x 22.5""

$DF - L$ GLB

Type.2   *Length*   12 ′

Similarly, the design for this length is performed.

Governing.Load.Combination

$D+L$ $C_D = 1.0$

LOADING			weigth		
wD	11.26	128	7.17	plf	
wL	33.33	400	plf		
wS	0.00	0.008	plf		

Duration Factor	ASD (D)	ASD (D+L)	ASD (D+S)	
CD	0.9	1	1.15	
Govern ASD ?	150.19	535.17	117.55	plf
		Governs		

Beam Dimensions					
b	2.125	in			
Laminations t	1.5	in	ro(pcf)	36.00	
n laminations	9		A(in^2)	28.69	
d	13.50	in	A(ft^2)	0.20	
			W(glulam)(plf)	7.17	

Beam Length					
L	144.00	12	ft	Mid span	
LA	0.00	0	ft	VOLUME(BC)ft^3	
LB	0.00	0	ft	2.39	

C.Factors

Adjustment Factors	Fb	Fv	F'c+	E
CM	1.00	1.00	1.00	1.00
Ct	1.00			
CL	1.00	Need to check for supports later		
CF (not for GLB)	1.00			
xyz	20			
k	1			
CV (only in GLB)	1.00	Note: Do not use CL and CV together		
Cfu	1.00			
Cr	1.00			
Cc	1.00			
Cf	1.00			

Bending

Design For Bending		
Mmax	115597	lbs-in
CD	1.00	
CM	1.00	
Ct	1.00	
CL	1.00	
CF	1.00	
CV	1.00	**Note:** Do not use CL and CV together
Cfu	1.00	
Cr	1.00	
Fb	2,400	psi
F'b	2,400	psi
b	2.125	in
d	13.50	in
Sx	64.55	in
fb	1,791	psi
check	**OK**	

Deflection

Deflection Limits				
	Calculated	Limits Deflection		
L	0.34	0.40	in	**OK**
D+L	0.42	0.60	in	**OK**

Shear

Design For Shear		
Vmax	3211	lbs
w	44.60	lbs/in
b	2.125	in
d	13.50	in
Area	28.69	in^2
V'	2609	lbs
CD	1.00	
CM	1.00	
Ct	1.00	
CH	1.00	
Fv	300	
F'v	300	psi
fv	168	psi
check	**OK**	

L/d = 11

Bearing

Design For Bearing		
R	3211	lbs
CM	1	
Ct	1	
Cb	1	Need to check supports later
Fc+	650	psi
F'c+	650	psi
b	2.125	in
lb	2.32	in
Use Base PL	CCO5 1/4-6 ECC	
R cap	15785	lbs
Check Plate	**OK**	
Use a lb	2.5	in
fv	604	psi
Cb	1.15	
F'c+	748	psi
check	**OK**	

CCO31/

ok

USE   "21/8″ x 13.5′″"

Summary of Purlin Design: "ROOF"

USE, Purlins as below:

$Type.1$ $L = 20\,'-4\,''$ USE "21/8″ x 28.5‴" GLB $DF-L$ $MC = 15$

$Type.2$ $L = 12\,'$ USE "21/8″ x 16.5‴" GLB $DF-L$ $MC = 15$

"Roof Purlin"

Note : Sub Purlin and Purlin have simply supported condition .

2 1/8" glu-lam joist 24F-V4 SP/SP - 36 pcf   - M.C=15%

D $16 \cdot psf \cdot 8 \cdot ft = 128 \ plf$

"self.weight:To Be Calculated"

L_r $16 \cdot psf \cdot 8 \cdot ft = 128 \ plf$

S $65 \cdot psf \cdot 8 \cdot ft = 520 \ plf$

$Type.1$ $Length$ $20'-4''$

$Type.2$ $Length$ $12'$

Note: The Purlins are categorised in two types that will fit two different lengths.
Design of beams are performed with the spread sheet. Bending,shear,bracing, and deflection are checked.

$Type.1$ $Length$ $20'-4''$

$A3-A2$ $(PLAN)$ $Check.Section.C$

$Type.2$ $Length$ $12'$

$A5-A4$ $(PLAN)$ $Check.Section.C$

Design   of   Type.1   Length   20 ′ − 4 ″

Governing.Load.Combination

$D + S$ $C_D := 1.15$

LOADING				weigth		
	wD	11.93	128	15.14	plf	
	wL	10.67	128	plf		
	wS	43.33	520	plf		
Duration Factor		ASD (D)	ASD (D+L)	ASD (D+S)		
	CD	0.9	1.25	1.15		
Govern ASD ?		159.05	216.91	576.64	plf	
				Governs		
Beam Dimensions						
	b	2.125	in			
Laminations t		1.5	in	ro(pcf)	36.00	
n laminations		19		A(in^2)	60.56	
d		28.50	in	A(ft^2)	0.42	
				W(glulam)(plf)	15.14	
Beam Length						
	L	243.96	20.33	ft	Mid span	

Bending

Design For Bending			
Mmax	741476	lbs-in	
CD	1.15		
CM	1.00		
Ct	1.00		
CL	1.00		
CF	1.00		
CV	1.00	**Note:** Do not use CL and CV together	
Cfu	1.00		
Cr	1.00		
Fb	2,400	psi	
F'b	2,760	psi	
b	2.125	in	
d	28.50	in	
Sx	287.67	in	
fb	2,578	psi	
check	OK		

Deflection

Deflection Limits				
	Calculated	Limits Deflection		
L	0.15	0.68	in	OK
D+L	0.23	1.02	in	OK

Shear

Design For Shear				
Vmax	7752	lbs		
w	22.60	lbs/in		
b	2.125	in		
d	28.50	in		
Area	60.56	in^2		
V'	7108	lbs		
CD	1.15			
CM	1.00			
Ct	1.00			
CH	1.00			
Fv	300			
F'v	345	psi		
fv	192	psi	L/d =	9
check	**OK**			

Bearing

Design For Bearing			
R	6741	lbs	
CM	1		
Ct	1		
Cb	1		Need to check supports later
Fc+	650	psi	
F'c+	650	psi	
b	2.125	in	CCO31/
lb	4.88	in	
Use Base PL	CCO5 1/4-6 ECC		
R cap	15785	lbs	
Check Plate	**OK**		
Use a lb	5	in	
fv	634	psi	
Cb	1.08		
F'c+	699	psi	
check	**OK**		

Adjustment.Factors

Adjustment Factors	Fb	Fv	F'c+	E
CM	1.00	1.00	1.00	1.00
Ct	1.00			
CL	1.00	Need to check for supports later		
CF (not for GLB)	1.00			
xyz	20			
k	1			
CV (only in GLB)	1.00	Note: Do not use CL and CV together		
Cfu	1.00			
Cr	1.00			
Cc	1.00			
Cf	1.00			

Therefore

USE   "2 1/8" x 28.5""

Type.2   *Length*   12 ′

$A5 - A4$ $(PLAN)$ *Check.Section.C*

Similarly, the design for this length is performed.

USE   "21/8″ x 16.5′″"

Governing.Load.Combination

$D + S$ $C_D = 1.15$

LOADING				weigth		
wD	11.40	128		8.77	plf
wL	10.67	128	plf		
wS	43.33	520	plf		
Duration Factor	ASD (D)	ASD (D+L)	ASD (D+S)		
CD	0.9	1.15	1.6		
Govern ASD ?	151.96	230.23	410.48	plf	
			Governs		
Beam Dimensions					
b	2.125	in			
Laminations t	1.5	in	ro(pcf)		36.00
n laminations	11		A(in^2)		35.06
d	16.50	in	A(ft^2)		0.24
			W(glulam)(plf)		8.77
Beam Length					
L	144.00	12	ft	Mid span	

Bending

Design For Bending		
Mmax	359424	lbs-in
CD	1.60	
CM	1.00	
Ct	1.00	
CL	1.00	
CF	1.00	
CV	1.00	**Note:** Do not use CL and CV together
Cfu	1.00	
Cr	1.00	
Fb	2,400	psi
F'b	3,840	psi
b	2.125	in
d	16.50	in
Sx	96.42	in
fb	3,728	psi
check	**OK**	

Deflection

Deflection Limits				
	Calculated	Limits Deflection		
L	0.09	0.40	in	**OK**
D+L	0.14	0.60	in	**OK**

Shear

Design For Shear		
Vmax	6305	lbs
w	22.06	lbs/in
b	2.125	in
d	16.50	in
Area	35.06	in^2
V'	5941	lbs
CD	1.60	
CM	1.00	
Ct	1.00	
CH	1.00	
Fv	300	
F'v	480	psi
fv	270	psi
check	**OK**	

L/d =	9

Bearing

Design For Bearing		
R	3941	lbs
CM	1	
Ct	1	
Cb	1	Need to check supports later
Fc+	650	psi
F'c+	650	psi
b	2.125	in	CCO31/
lb	2.85	in
Use Base PL	CCO5 1/4-6 ECC	
R cap	15785	lbs
Check Plate	**OK**	
Use a lb	6.75	in
fv	275	psi
Cb	1.00	
F'c+	650	psi
check	**OK**	

Therefore

USE   "21/8″ x 16.5″"

Summary of Sub-Purlin Design: "FLOOR"

$$USE \qquad DF-L \qquad 4\ x8 \quad @4' \qquad MC=15$$

Use for all of the Sub-Purlins in the First Floor.

"Floor Sub−Purlin"

Location, all of Sub-Purlins of First Floor.

Length   8 *ft*

Loads

   D $16 \cdot psf \cdot 4 \cdot ft = 64 \ plf$

        "self.weight:To Be Calculated"

   L $50 \cdot psf \cdot 4 \cdot ft = 200 \ plf$

GOVERNING.LOAD.COMB

        $D + L$ $C_D = 1.00$

                                        USE $DF - L$ 4 *x*8 @4′ $MC = 15$

GIVEN PROBLEM STATEMENT			
Wood Species Southern		S	
Beam Type 24-V1 SP/Sp			
MC limit	16	Note: 16 for GLB 19 Solid Sawn	
Moisture	15		
Density of GLB	36		
Deflection Limits			
Ceiling type	Plaster	P	
Deflection L	360		
Deflection D+L	240		
LOADING		weigth	
wD	5.86	64	6.34 plf
wL	16.67	200 plf	
wS	0.00	0.008 plf	
Duration Factor	ASD (D)	ASD (D+L) ASD (D+S)	
CD	0.9	1 1.15	
Govern ASD ?	78.16	270.34 61.18	plf
		Governs	
Beam Dimensions			
b	3.500	in	
Laminations t	7.25	in	ro(pcf) 36.00
n laminations	1		A(in^2) 25.38
d	7.25	in	A(ft^2) 0.18
			W(glulam)(plf) 6.34
Beam Length			
L	96.00	8 ft	Mid span
LA	0.00	0 ft	VOLUME(BC)ft^3
LB	0.00	0 ft	1.41

Bending   &   Deflection

	Design For Bending		
Mmax	25953	lbs-in	
CD	1.00		
CM	1.00		
Ct	1.00		
CL	1.00		
CF	1.00		
CV	1.00	**Note:** Do not use CL and CV together	
Cfu	1.00		
Cr	1.00		
Fb	1,500	psi	
F'b	1,500	psi	
b	3.500	in	
d	7.25	in	
Sx	30.66	in	
fb	846	psi	
check	**OK**		

	Deflection Limits				
	Calculated	Limits Deflection			
L	0.13	0.27	in		**OK**
D+L	0.17	0.40	in		**OK**

Shear

	Design For Shear		
Vmax	1081	lbs	
w	22.53	lbs/in	
b	3.500	in	
d	7.25	in	
Area	25.38	in^2	
V'	918	lbs	
CD	1.00		
CM	1.00		
Ct	1.00		
CH	1.00		
Fv	300		
F'v	300	psi	
fv	64	psi	L/d = 13
check	**OK**		

Bearing

	Design For Bearing		
R	1081	lbs	
CM	1		
Ct	1		
Cb	1	Need to check supports later	
Fc+	650	psi	
F'c+	650	psi	
b	3.500	in	CCO31/
lb	0.48	in	
Use Base PL	CCO5 1/4-6 ECC		
R cap	15785	lbs	
Check Plate	**OK**		
Use a lb	1	in	
fv	309	psi	
Cb	1.38		
F'c+	894	psi	
check	**OK**		

C_L

Design for bracing
lu-top	48.00	4	ft	Assume purlines/joits at 4ft on center.
lu-bottom	96.00	8	ft	max of l1 and l2 times 2
b	3.500	in		
d	7.25	in		

Design for stability
E'y	1,700,000	psi
	GLB	
Kbe	0.610	
lu/d top	6.62	
lu/d botto	13.24	

$$l_e = \begin{cases} 2.06\, l_u & l_u/d < 7 \\ 1.63\, l_u + 3d & 7 \le l_u/d \le 14.3 \\ 1.84\, l_u & l_u/d > 14.3 \end{cases}$$

le top	98.88	
le bottom	178.23	
Rb top	7.65	OK
Rb botton	10.27	OK

$$R_B = \sqrt{\frac{l_e d}{b^2}} \le 50$$

FbE-top	17720	psi
FbE-bottc	9831	psi

Fb*	1500	psi

Remove the CV since CL is not to be used with CL

Q-top	11.81
Q-bottom	6.55

$$F_{bE} = \frac{K_{bE} E'_y}{R_B^2}$$

CL top	1.00
CL bot	0.99

$$Q = \frac{F_{bE}}{Fb^*}$$

F'b top	1493	psi
F'b bot	1487	psi

$$C_L = \left[\left(\frac{1+Q}{1.9} \right) - \sqrt{ \left(\frac{1+Q}{1.9} \right)^2 - \frac{Q}{0.95} } \right] \le 1.0$$

fb	846	psi
	OK	
	OK	

ok

Adjustment.Factors

Adjustment Factors	Fb	Fv	F'c+	E
CM	1.00	1.00	1.00	1.00
Ct	1.00			
CL	1.00	Need to check for supports later		
CF (not for GLB)	1.00			
xyz	20			
k	1			
CV (only in GLB)	1.00	Note: Do not use CL and CV together		
Cfu	1.00			
Cr	1.00			
Cc	1.00			
Cf	1.00			

Therefore, this size is adequate.

Summary of Sub-Purlin Design: "Roof"

Length   8 *ft*

USE $DF-L$ $4\ x10$ $@4'$ $MC=15$

Use for all of the Sub-Purlins in the Roof.

"Roof Sub−Purlin"

Location, all of Sub-Purlins of First Floor.

Length 　　　 *8 ft*

Loads

Dead (Type)	*Load (psf)*
"Built up flooring"	4
"Sub−flooring"	1.5
"R−19 Floor Insullation"	1.5
"1/8 T&G Pine Wood Ceiling"	4
"Misc + Light Fixtures"	5
"Total"	16

Roof.Live	*Load (psf)*
"Roof Live Load"	16

Snow	*Load (psf)*
"Snow Load"	65

Note: The load is converted to horizontal equivalent

Governing.Load.Combination 　　　 $D + S$

$$C_D := 1.15$$

USE   DF − L   4 x10   @4′   MC = 15

Governing.Load.Combination

GIVEN PROBLEM STATEMENT				
Wood Species Southern		S		
Beam Type 24-V1 SP/Sp				
MC limit	16	Note: 16 for GLB 19 Solid Sawn		
Moisture	15			
Density of GLB	36			
Deflection Limits				
Ceiling type	Plaster	P		
Deflection L	360			
Deflection D+L	240	4x10		
LOADING			weigth	
wD	5.96	64	7.52	plf
wL	5.33	64	plf	
wS	21.67	260	plf	
Duration Factor	ASD (D)	ASD (D+L)	ASD (D+S)	
CD	0.9	1.15	1.15	
Govern ASD ?	79.46	117.84	288.27	plf
			Governs	
Beam Dimensions				
b	3.250	in		
Laminations t	9.3	in		
n laminations	1			
d	9.25	in		

ro(pcf)	36.00
A(in^2)	30.06
A(ft^2)	0.21
W(glulam)(plf)	7.52

Beam Length				
L	96.00	8	ft	
LA	0.00	0	ft	
LB	0.00	0	ft	
al	0.000		L=	8.00

Mid span	
VOLUME(BC)ft^3	
1.67	

Bending

Design For Bending		
Mmax	57408	lbs-in
CD	1.15	
CM	1.00	
Ct	1.00	
CL	1.00	
CF	1.00	
CV	1.00	**Note:** Do not use CL and CV together
Cfu	1.00	
Cr	1.00	
Fb	1,500	psi
F'b	1,725	psi
b	3.250	in
d	9.25	in
Sx	46.35	in
fb	1,239	psi
check	OK	

Deflection

Deflection Limits				
	Calculated	Limits Deflection		
L	0.03	0.27	in	OK
D+L	0.05	0.40	in	OK

Shear

Design For Shear				
Vmax	1525	lbs		
w	11.29	lbs/in		
b	3.250	in		
d	9.25	in		
Area	30.06	in^2		
V'	1421	lbs		
CD	1.15			
CM	1.00			
Ct	1.00			
CH	1.00			
Fv	95			
F'v	109	psi		
fv	76	psi	L/d =	10
check	OK			

Bearing

Design For Bearing			
R	1326	lbs	
CM	1		
Ct	1		
Cb	1	Need to check supports later	
Fc+	625	psi	
F'c+	625	psi	
b	3.250	in	CCO31/
lb	0.65	in	
Use Base PL	CCO5 1/4-6 ECC		
R cap	15785	lbs	
Check Plate	**OK**		
Use a lb	1	in	
fv	408	psi	
Cb	1.38		
F'c+	859	psi	
check	**OK**		

Bracing.C_L

Design for bracing

lu-top	48.00	4	ft	Assume purlines/joits at 4ft on center.
lu-bottom	96.00	8	ft	max of l1 and l2 times 2
b	3.250	in		
d	9.25	in		

Design for stability

E'y	1,900,000	psi	
	GLB		
Kbe	0.610		
lu/d top	5.19		
lu/d bottor	10.38		

$$l_e = \begin{cases} 2.06\, l_u & l_u/d < 7 \\ 1.63\, l_u + 3d & 7 \le l_u/d \le 14.3 \\ 1.84\, l_u & l_u/d > 14.3 \end{cases}$$

le top	98.88	
le bottom	184.23	
Rb top	9.31	OK
Rb bottom	12.70	OK

$$R_B = \sqrt{\frac{l_e d}{b^2}} \le 50$$

FbE-top	13384	psi
FbE-botto	7184	psi
Fb*	1725	psi	Remove the CV since CL is not to be used with CL

$$F_{bE} = \frac{K_{bE} E'_y}{R_B^2}$$

Q-top	7.76	
Q-bottom	4.16	

$$Q = \frac{F_{bE}}{F b^+}$$

CL top	0.99	
CL bot	0.98	

F'b top	1712	psi
F'b bot	1699	psi

$$C_L = \left[\left(\frac{1+Q}{1.9}\right) - \sqrt{\left(\frac{1+Q}{1.9}\right)^2 - \frac{Q}{0.95}}\right] \le 1.0$$

fb	1239	psi
	OK	
	OK	

Adjustment.Factors

Adjustment Factors	Fb	Fv	F'c+	E
CM	1.00	1.00	1.00	1.00
Ct	1.00			
CL	1.00	Need to check for supports later		
CF (not for GLB)	1.00			
xyz	20			
k	1			
CV (only in GLB)	1.00	Note: Do not use CL and CV together		
Cfu	1.00			
Cr	1.00			
Cc	1.00			
Cf	1.00			

Therefore, this size is adequate.

Summary of Design for Header: **" First Floor "**

USE 6x10 DF-L Select Structural for all of the headers of first floor (Floor).
This size meet the requirement for Chord force and Strut Force in addition to all
of loading combinations.

Designinig a Header: *Floor*

Lateral   force $65.48 \cdot kip$

direction.1 − 1

$$W_1 := \frac{\dfrac{65.48 \cdot kip}{2}}{188 \cdot ft} = 0.174 \ \frac{kip}{ft}$$

$$M_1 := \frac{W_1 \cdot (188 \cdot ft)^2}{8} = 769.39 \ kip \cdot ft$$

$$C_1 := \frac{M_1}{66 \cdot ft} = 11.657 \ kip \qquad chord.force.in.short.side$$

direction.2 − 2

$$W_2 := \frac{\dfrac{65.48 \cdot kip}{2}}{66 \cdot ft} = 0.496 \ \frac{kip}{ft}$$

$$M_2 := \frac{W_2 \cdot (66 \cdot ft)^2}{8} = 270.105 \ kip \cdot ft$$

$$C_1 := \frac{M_2}{188 \cdot ft} = 1.437 \ kip \qquad chord.force.in.long.side$$

Chord force in 1-1 directionis governing.
The longest headerin thar side is 18'-9" long.

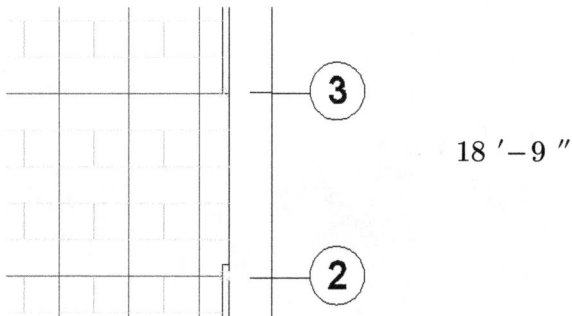

$18' - 9''$

"header of H−23 is considered "

Wall.Dead.Load $16 \cdot psf$

Floor Dead Load w/o GLB $22 \cdot psf$ *tributary.Area* $2 \cdot ft \cdot 18.75 \cdot ft = 37.5 \ ft^2$

Live.Load $50 \cdot psf$

Governing.Load.Comb.for.bending

$D + L$ $C_D := 1.00$

Governing.Load.Comb.for.bending&compression

$D + L + 0.7 \ E$ $C_D := 1.6$

NOTE: NO GLB is being seated on header.
$L := 18.75 \cdot ft$

Member Properties:

6x member
DF-L Select structural
MC<15%

Bending

D $22 \cdot psf \cdot 2 \cdot ft = 44 \ plf$

L $50 \cdot psf \cdot 2 \cdot ft = 100 \ plf$

Try   6 x10   member

Design For Bending		
Mmax	82826	lbs-in
CD	1.00	
CM	1.00	
Ct	1.00	
CL	1.00	
CF	1.00	
CV	1.00	
Cfu	1.00	
Cr	1.00	
Fb	1,500	psi
F'b	1,500	psi
b	5.500	in
d	9.50	in
Sx	82.73	in
fb	1,001	psi
check	**OK**	

Design For Shear		
Vmax	1472	lbs
w	13.09	lbs/in
b	5.500	in
d	9.50	in
Area	52.25	in^2
V"	1348	lbs
CD	1.00	
CM	1.00	
Ct	1.00	
CH	1.00	
Fv	95	
F'v	95	psi
fv	42	psi
check	**OK**	

L/d = 24

Deflection Limits				
	Calculated	Limits	Deflection	
L	0.58	0.63	in	**OK**
D+L	0.80	0.94	in	**OK**

Design for bracing

lu-top	48.00	4	ft	Assume purlines/joits at 4ft on center.
lu-bottom	48.00	4	ft	max of l1 and l2 times 2
b	5.500	in		
d	9.50	in		
Design for stability				
E'y	1,900,000	psi		
	GLB			
Kbe	0.610			
lu/d top	5.05			
lu/d bottor	5.05			

$$l_e = \begin{cases} 2.06\, l_u & l_u/d < 7 \\ 1.63\, l_u + 3d & 7 \le l_u/d \le 14.3 \\ 1.84\, l_u & l_u/d > 14.3 \end{cases}$$

le top	98.88	
le bottom	98.88	
Rb top	5.57	**OK**
Rb bottor	5.57	**OK**

$$R_B = \sqrt{\frac{l_e d}{b^2}} \le 50$$

FbE-top	37323	psi
FbE-bottc	37323	psi
Fb*	1500	psi

Remove the CV since CL is not to be used with CL

Q-top	24.88	
Q-bottom	24.88	

$$F_{bE} = \frac{K_{bE} E'_y}{R_B^2}$$

CL top	1.00	
CL bot	1.00	

$$Q = \frac{F_{bE}}{Fb^*}$$

F'b top	1497	psi
F'b bot	1497	psi
fb	1001	psi
	OK	
	OK	

$$C_L = \left[\left(\frac{1+Q}{1.9}\right) - \sqrt{\left(\frac{1+Q}{1.9}\right)^2 - \frac{Q}{0.95}}\right] \le 1.0$$

ok

Compression

P	8,160	lbs	
d (in.)	9.5	b (in.)	5.5
	compression		
Fc	1300	psi	
CD	1.6		
CF	1		
Cr	1		
Fc*	2080	psi	w/o Cp
	col capacity		
ke	1.0		ft
luy	225.0	L	18.75
le/d	23.7	OK	
Emin	620000.0	psi	
FcE	908.5	ok	
Fc*	2080.0	psi	
Qy	0.4	psi	
c	0.8		
Q	0.4		
Cp	0.4		
F'c	806.4		
fc	156.2		
condition	OK		

Compression & Bending

Compression and bending		
fc	156.17	psi
F'c	806.42	psi
F'b	1500.00	psi
FcE	908.54	psi
fb1	1001.17	
Allowable stress ratio	0.84	1
condition	OK	

USE 6x10 DF-L Select Structural for Floor Headers

Bending and Tension:

Note: Since the members also experience bending and tension, current design has to be checked against this condition.

In other words header will experience a distributed load and tension force.(P)
Note: 0.7 coefficient has been applied to p.

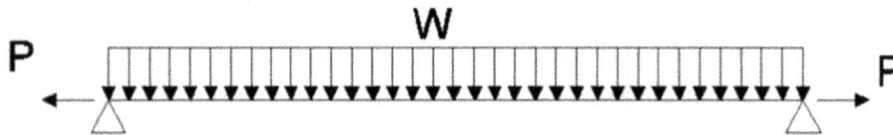

Tension and bending		
ft(capacity)	1000	psi
F't	1600.000	psi
F'b	1,500	psi
fb1	1,001	psi
A(net)	52	in^2
ft	156.17	psi
Allowable stress ratio	0.77	1
condition	OK	

Therefore,

USE 6x10 DF-L Select Structural for all of the headers of first floor(Floor).

Note: for the openings that are located on the 2-2 direction (long) , a post is required to install at the middle of 24 ft length in order to reduce the lngth of each span from 24 ft to 12 ft to have a satisfactory design for all of members.

Checking adequacy of headers against Strut force:

Since, most probably the design of headers will govern with chord forces. Therefore, initially design is based on chord force and subsequently the designwill be checked for strut forces.

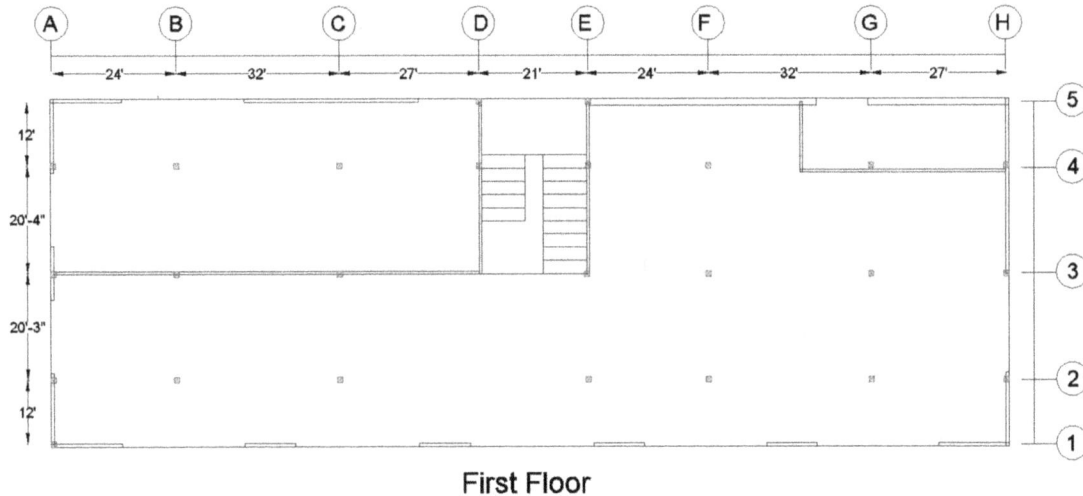

First Floor

First floor is going to experience the governing strut force.

$$L_{swr} := 47.25 \cdot ft \qquad L_{swf} := 47.25 \cdot ft \qquad op1 := 18.75 \cdot ft$$

$$S_{Wu} := \left(\frac{\frac{116.92 \cdot kip}{2}}{L_{swr}} \right) = 1.237 \, \frac{kip}{ft}$$

$$S_{Wf} := S_{Wu} + \left(\frac{\frac{65.48 \cdot kip}{2}}{L_{swf}} \right) = 1.93 \, \frac{kip}{ft}$$

$$S_{floor} := \left(\frac{(116.92 + 65.48) \cdot kip}{2} \right) \cdot \left(\frac{1}{66 \cdot ft} \right) = 1.382 \, \frac{kip}{ft}$$

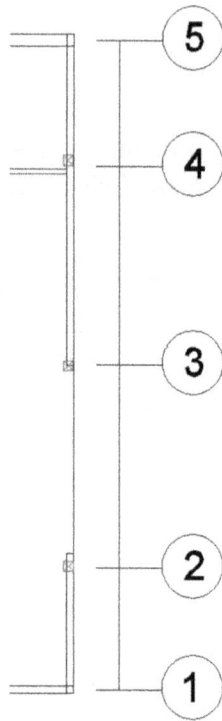

Summary of strut force for short direction (1-1) L=66ft

Floor

The.Force.diagram

X	0	14	32.75	66
Force(kip)	0	7.7	-18.175	0

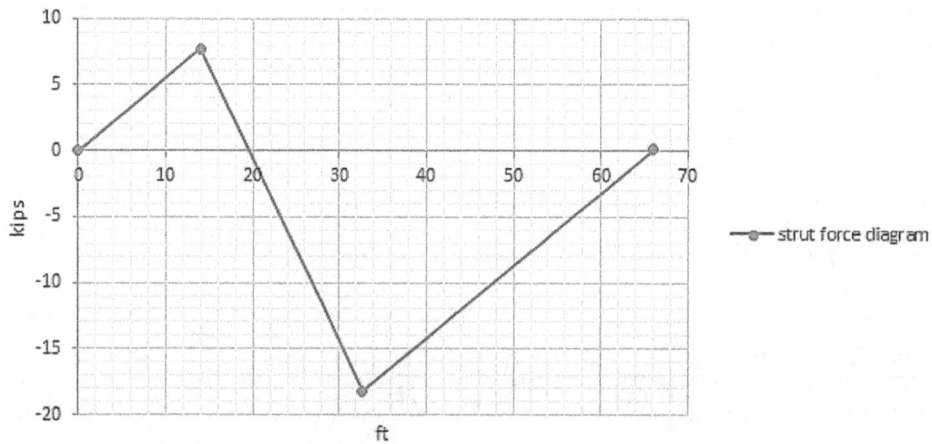

Approved. According to this force the 6x10 member is adequate .Therefore, Use 6x10 DF-L Select.Structural for all of headers in the floor.

Summary of Design for Header: " Roof"

USE 6x12 DF-L Select Structural for all of the headers of Roof.
This size meets the requirement for Chord force and Strut Force in addition to all
of loading combinations.

Designinig a Header: *Roof*

Most probably, The Chord force is going to be a governing case for headers. Therefore, for each floor, the chord force is calculated, and the beam is going to be designed as a beam column(combination of bending and compression), this opinion will be verified.

1

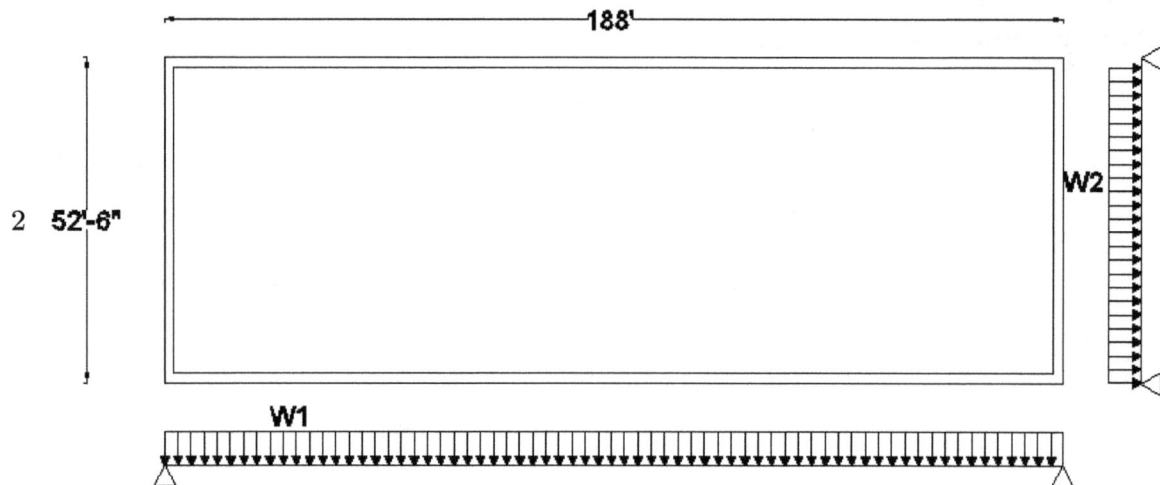

1

Roof

Lateral   force $116.93 \cdot kip$

direction.$1-1$

$$W_1 := \frac{\frac{116.93 \cdot kip}{2}}{188 \cdot ft} = 0.311 \; \frac{kip}{ft}$$

$$M_1 := \frac{W_1 \cdot (188 \cdot ft)^2}{8} = 1373.928 \; kip \cdot ft$$

$$C_1 := \frac{M_1}{52.5 \cdot ft} = 26.17 \; kip \qquad chord.force.in.short.side$$

$direction.2-2$

$$W_2 := \frac{\frac{116.93 \cdot kip}{2}}{52.5 \cdot ft} = 1.114 \, \frac{kip}{ft}$$

$$M_2 := \frac{W_2 \cdot (52.5 \cdot ft)^2}{8} = 383.677 \, kip \cdot ft$$

$$C_1 := \frac{M_2}{188 \cdot ft} = 2.041 \, kip \qquad chord.force.in.long.side$$

Chord force in 1-1 directionis governing.
The longest headerin thar side is 18'-9" long.

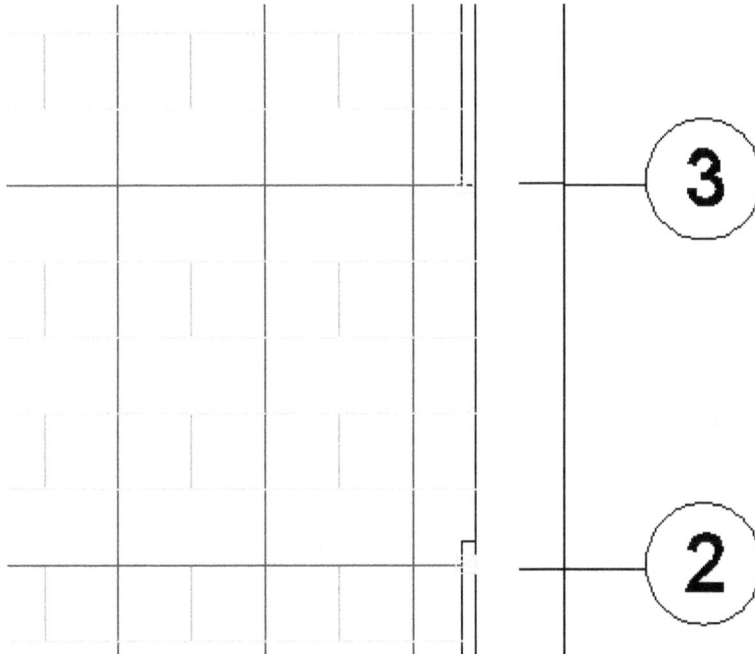

"header of H−23 is considered "

$Wall.Dead.Load \qquad 16 \cdot psf$

$Roof \; Dead \; Load \; w/o \; GLB \qquad 22 \cdot psf \qquad tributary.Area \quad 2 \cdot ft \cdot 18.75 \cdot ft = 37.5 \, ft^2$

$Snow.Load \qquad 65 \cdot psf$

Roof.Live $16 \cdot psf$

Governing.Load.Comb.for.bending

   $D + S$ $C_D := 1.15$

Governing.Load.Comb.for.bending&compression

   $D + S + 0.7\ E$ $C_D := 1.6$

NOTE: NO GLB is being seated on header.
   $L := 18.75 \cdot ft$

Member Properties:

6x member
DF-L Select structural
MC<15%

Bending

D $22 \cdot psf \cdot 2 \cdot ft = 44\ plf$

S $65 \cdot psf \cdot 2 \cdot ft = 130\ plf$

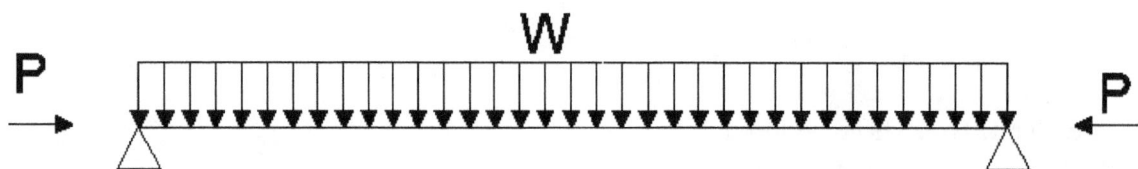

Note: Since when there is a combined bending and compression , 0.7 factor has to be multiplied by chord force .

Try   6 *x12*   *member*

Check bending,deflection, and shear is satisfactory with help of given spreadsheet. OK

Compression

P	26,170	lbs		
d (in.)	11.5	b (in.)		5.5
	compression			
Fc	1300			
CD	1.6			
CF	1			
Cr	1			
Fc*	2080	psi		w/o Cp
	col capacity			
ke	1.0			ft
luy	225.0	L		18.75
le/d	19.6	OK		
Emin	620000.0	psi		
FcE	1331.4	ok		
Fc*	2080.0	psi		
Qy	0.6	psi		
c	0.8			
Q	0.6			
Cp	0.5			
F'c	1090.8			
fc	413.8			
condition	OK			

Compression&Bending

Compression and bending		
fc	413.75	psi
F'c	1090.79	psi
F'b	2400.00	psi
FcE	1331.36	psi
fb1	717.98	
Allowable stress ratio	0.58	1
condition	OK	

OK

USE 6x12 DF-L Select Structural for Roof Headers

Bending and Tension:

Note: Since the members also experience bending and tension, currunt design has to be checked against this condition.

In other words header will experience a distributed load and tension force.(P)
Note: 0.7 coefficient has been applied to p.

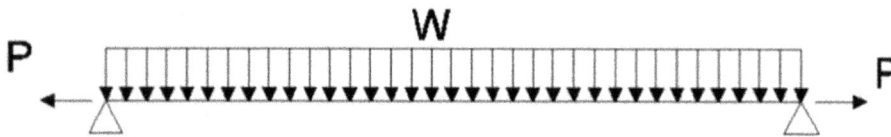

Tension and bending		
ft(capacity)	1000	psi
F't	1600.000	psi
F'b	2,400	psi
fb1	718	psi
A(net)	63	in^2
ft	413.75	psi
Allowable stress ratio	0.56	1
condition	OK	

Therefore,

USE 6x12 DF-L Select Structural for all of the headers of second floor(roof).

Note: Strut force is checked for higher force and smaller section which was appropriate (first floor) .Therefore, this size is valid against any strut force.

Front Balcony Design Summary:　　　(Floor)

Two alternatives are presented for dealing with the balcony and difference between loading the roof on the vicinity of the balcony.

Alternative.1

According to the load combination, the Load combination that has dead load and Snow load are going to be governing. Therefore, the performed design for the below the balcony is not adequate.
It means that the performed analysis and design for the roof fully satisfy the requirement for the floor. Therefore, Use the same design for the roof as purlin, Sub-Purlin, and GLBs for the floor to be safe. The different size between floor and roof is very small.So, this alternative will work fine for dealing with balcony.

Alternative.2

The other alternative is extending the roof to the edge of the balcony, in this case, we can use the current design for the floor based on the fact that snow load is not governing and design is safe.The drawings for the second alternative is on the next page.In this case, there is no need to change the design for the floor and roof, and it is possible to use different sizes for floor and roof as being calculated.

Note: It is recommended to discuss the opinion of client and architect for applying each alternative.

Front Balcony Design :   (Floor)

Two alternatives are presented for dealing with the balcony and difference between loading the roof on the vicinity of the balcony.

Alternative.1

According to the load combination, the Load combination that has dead load and Snow load are going to be governing. Therefore, the performed design for the below the balcony is not adequate.

It means that the performed analysis and design for the roof fully satisfy the requirement for the floor. Therefore, Use the same design for the roof as purlin, Sub-Purlin, and GLBs for the floor to be safe. The different size between floor and roof is very small.So, this alternative will work fine for dealing with balcony.

Therefore, use the same members of the roof for the floor. In this case, the design will be safe for snow load combination.

Alternative.2

The other alternative is extending the roof to the edge of the balcony, in this case, we can use the current design for the floor based on the fact that snow load is not governing and design is safe. The drawings for the second alternative is on the next page. In this case, there is no need to change the design for the floor and roof, and it is possible to use different sizes for floor and roof as being calculated.

Section View point:

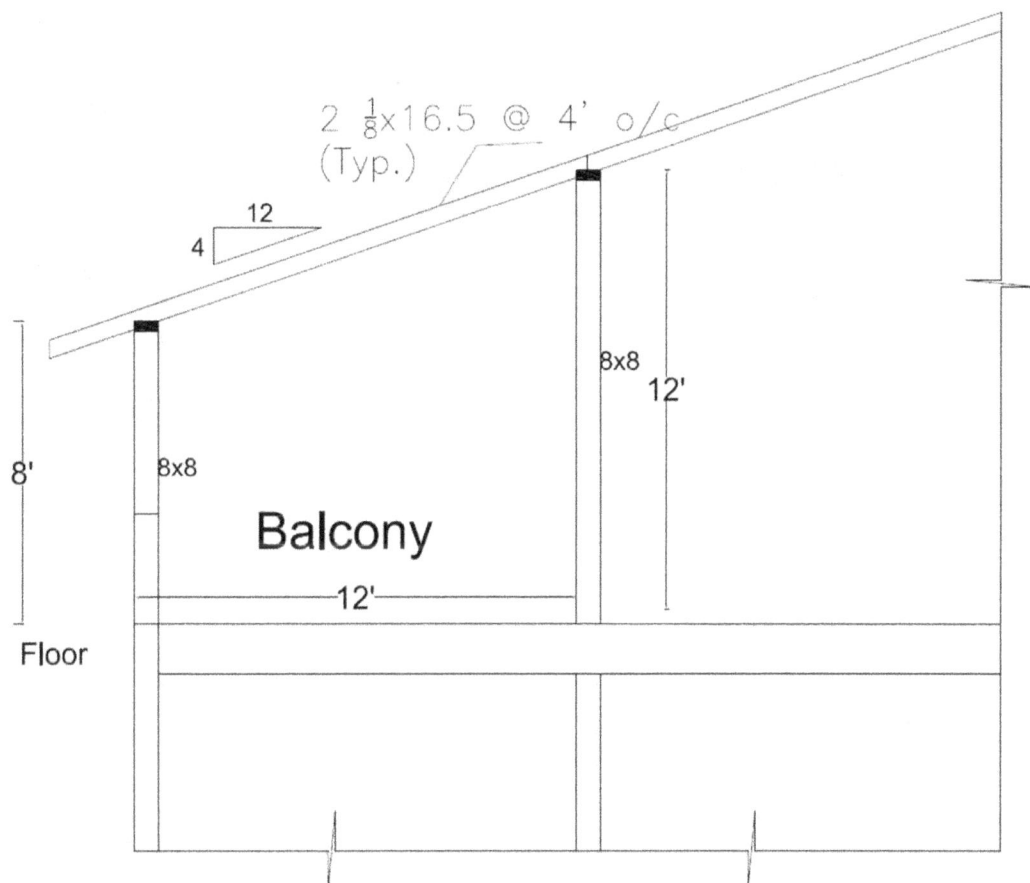

Extending the roof to the front of the balcony.

COVER PAGE : FLOOR NAILING DESIGN

Design of Nailing Schedule: (Floor)

Summary of floor nailing Design:

PLAN

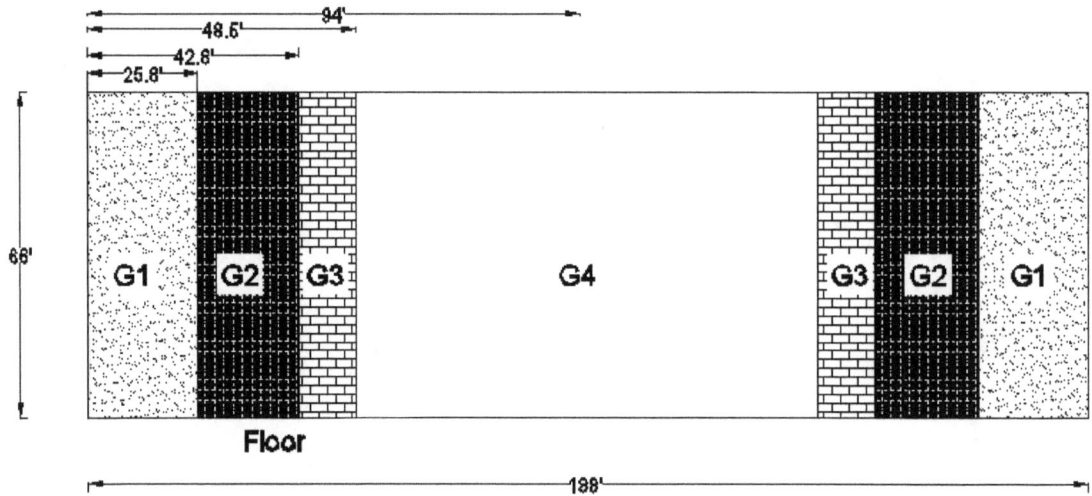

Floor

Group Specfications :

Group	Lable	Capcity Vxi	Plywood t	Min framing b	Continuous edges	Other Edges	Blocked edges	Capacity(plf)	Nail size	APA		
1	X0	530	3/8	2x	2 1/2	4	Y	530	8d	4	2 1/2	2x
2	X1	360	3/8	2x	4	6	Y	360	8d	6	4	2x
3	X2	270	3/8	2x	6	6	Y	270	8d	6	6	2x
4	X3	240	3/8	2x	6	0	N	240	8d	0	6	2x

Design of Nailing Schedule: (Floor)

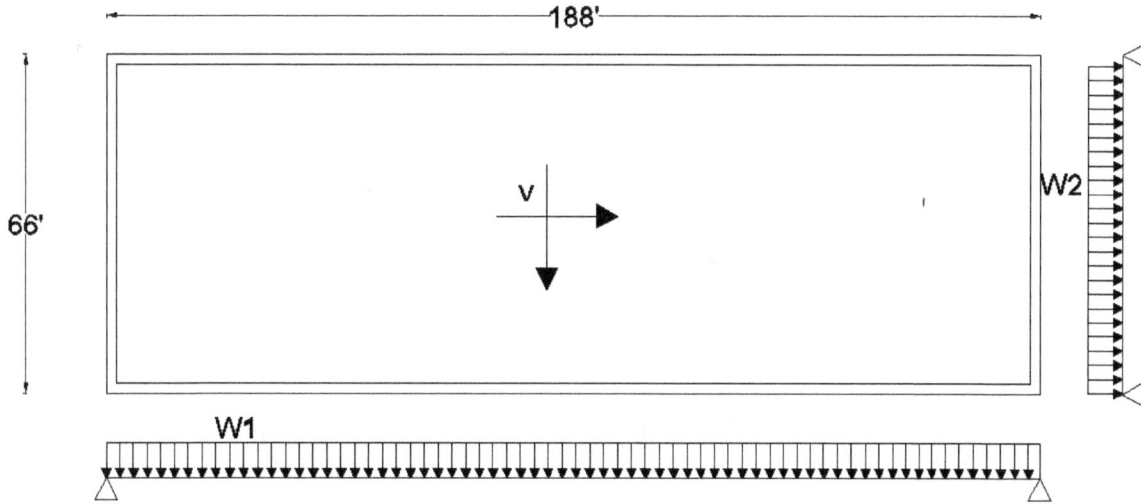

$$V_f := 65.48 \cdot kip \qquad L_1 := 188 \cdot ft \qquad L_2 := 66 \cdot ft$$

$$R_1 := \frac{V_f}{2} = 32.74 \ kip$$

Plywood layout is parallel to the longitudinal direction.(It is important for design)

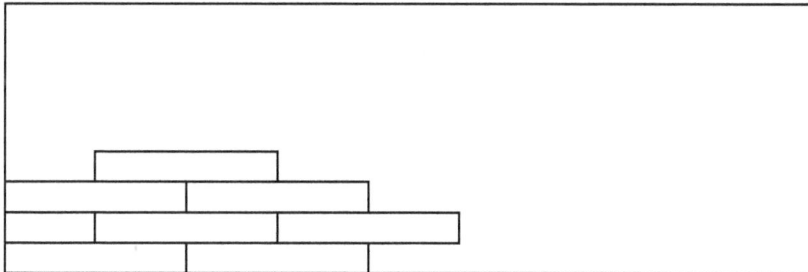

ALLOWABLE SHEAR (POUNDS PER FOOT) FOR HORIZONTAL APA PANEL DIAPHRAGMS WITH FRAMING OF DOUGLAS-FIR, LARCH OR SOUTHERN PINE[M] FOR WIND OR SEISMIC LOADING[M] (See also IBC Table 2306.3.1)

					Blocked Diaphragms				Unblocked Diaphragms	
					Nail Spacing (in.) at diaphragm boundaries (all cases), at continuous panel edges parallel to load (Cases 3 & 4), and at all panel edges (Cases 5 & 6)[M]				Nails Spaced 6" max. at Supported Edges[M]	
				Minimum Nominal Width of Framing Member at Adjoining Panel Edges and Boundaries (in.)	6	4	2-1/2[M]	2[M]	Case 1 (No unblocked edges or continuous joints parallel to load)	All other configurations (Cases 2, 3, 4, 5 & 6)
Panel Grade	Common Nail Size[F]	Minimum Nail Penetration in Framing (in.)	Minimum Nominal Panel Thickness (in.)		Nail Spacing (in.) at other panel edges (Cases 1, 2, 3 & 4)[M]					
					6	6	4	3		
APA STRUCTURAL I grades	6d[M] (0.113" dia.)	1-1/4	5/16	2	185	250	375	420	165	125
				3	210	280	420	475	185	140
	8d (0.131" dia.)	1-3/8	3/8	2	270	360	530	600	240	180
				3	300	400	600	675	265	200

Short Direction

The.Demand

Calculations	NS	Short direction	
longer edge	L	188.0	ft
Total shear of floor	V	65.5	kip
	W	348.3	plf
	R	32740.0	V/2 lbf
shorter edge	H	66.0	ft
flow(DEMAND)	Shear	496.1	plf

Design

Group	Lable	Capcity Vxi
1	X0	530
2	X1	360
3	X2	270
no blck case 1 4	X3	240

Group	Lable	Capcity Vxi	Plywood t	Min framing	Continuous edges	Other Edges	Blocked edges	Capacity(plf)	Nail size			APA	
1	X0	530	3/8	2x	2 1/2	4	Y	530	8d	4		2 1/2	2x
2	X1	360	3/8	2x	4	6	Y	360	8d	6	4		2x
3	X2	270	3/8	2x	6	6	Y	270	8d	6	6		2x
4	X3	240	3/8	2x	6	0	N	240	8d	0	6		2x

USE

Group	from X=	to X=
1	0	25.8
2	25.8	42.8
3	42.8	48.5
4	48.5	94.0

symmetric for both sides

Long Direction

Long   span

Calculations	NS	Short direction	
shorter edge L	L	66.0	ft
Total shear of floor	V	65.5	kip
	W	992.1	plf
	R	32740.0	V/2 lbf
Longer edge	H	188.0	ft
flow(DEMAND)	Shear	174.1	plf

It can be noticed that no blocking for case 2 is also workig.

Nailing layout:

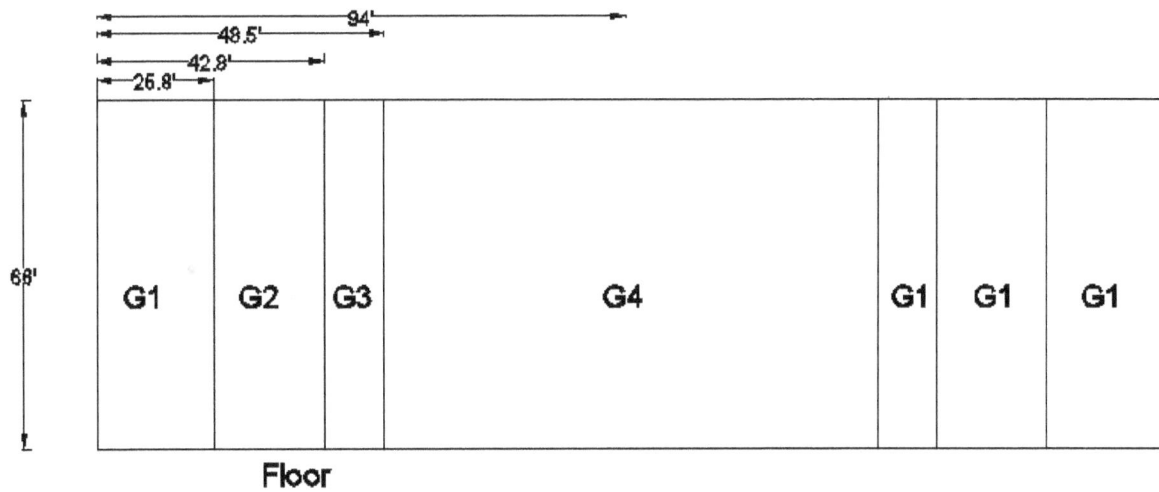

Floor

COVER PAGE : ROOF NAILING DESIGN

Design of Nailing Schedule: (Roof)

Summary of Roof nailing Design:

PLAN

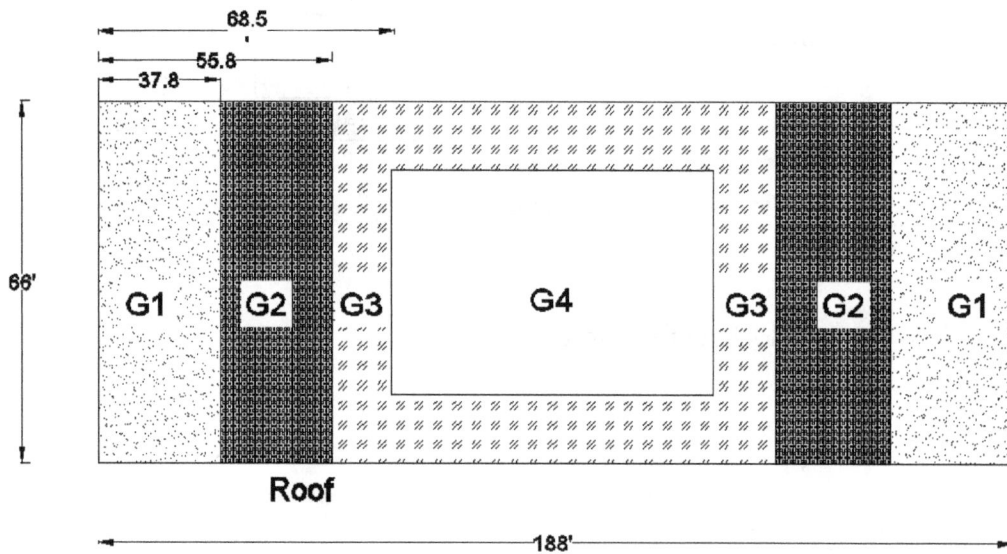

Roof

Group Specfications :

Group	Lable	Capcity Vxi	Plywood t	Min framing b	Continuous edges	Other Edges	Blocked edges	Capacity(plf)	Nail size	APA		
1	X0	1290	15/32	2(2x)	2 1/2	3	Y	1290	10d	3	2 1/2	2(2x)
2	X1	530	3/8	2x	2 1/2	4	Y	530	8d	4	2 1/2	2x
3	X2	360	3/8	2x	6	4	Y	360	8d	4	6	2x
4	X3	240	3/8	2x	6	0	N	240	8d	0	6	2x

no blck case 1

Design of Nailing Schedule: (Roof)

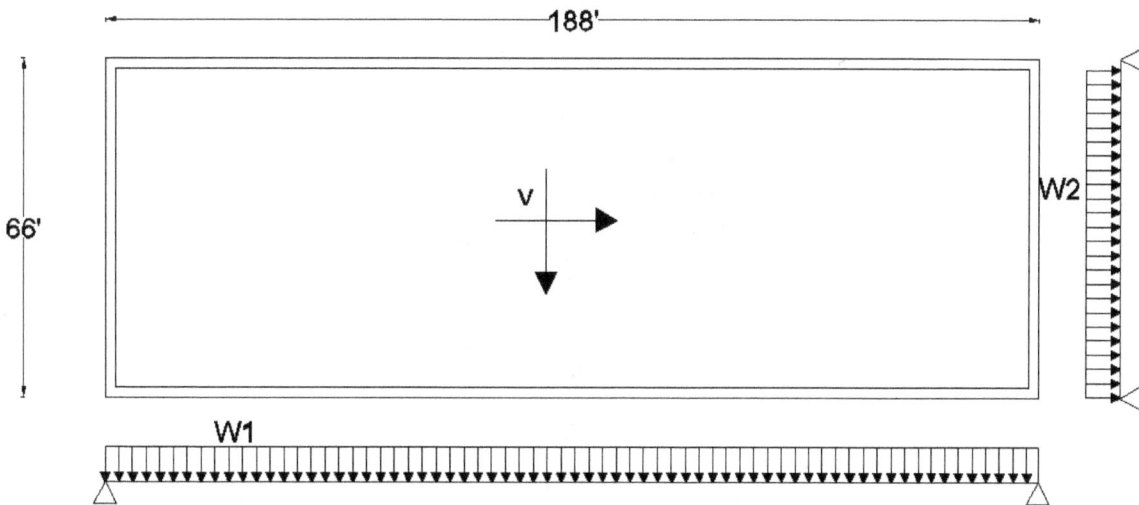

$$V_r := 116.93 \cdot kip \qquad L_1 := 188 \cdot ft \qquad L_2 := 66 \cdot ft$$

$$R_1 := \frac{V_r}{2} = 58.465 \; kip$$

Plywood layout is parallel to the longitudinal direction.(It is important for design)

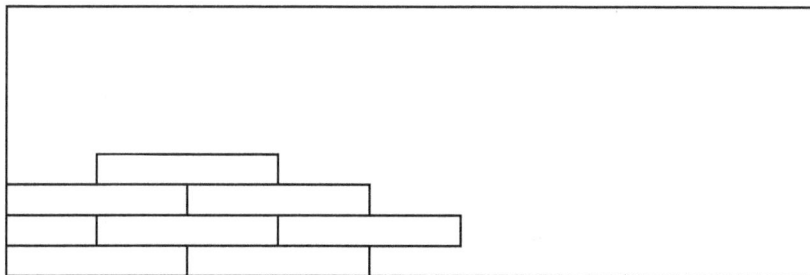

ALLOWABLE SHEAR (POUNDS PER FOOT) FOR HORIZONTAL APA PANEL DIAPHRAGMS WITH FRAMING OF DOUGLAS-FIR, LARCH OR SOUTHERN PINE[h] FOR WIND OR SEISMIC LOADING[a] (See also IBC Table 2306.3.1)

					Blocked Diaphragms				Unblocked Diaphragms	
					Nail Spacing (in.) at diaphragm boundaries (all cases), at continuous panel edges parallel to load (Cases 3 & 4), and at all panel edges (Cases 5 & 6)[b]				Nails Spaced 6" max. at Supported Edges[b]	
				Minimum Nominal Width of Framing Member at Adjoining Panel Edges and Boundaries (in.)	6	4	2-1/2[b]	2[b]	Case 1 (No unblocked edges or continuous joints parallel to load)	All other configurations (Cases 2, 3, 4, 5 & 6)
		Minimum Nail Penetration in Framing (in.)	Minimum Nominal Panel Thickness (in.)		Nail Spacing (in.) at other panel edges (Cases 1, 2, 3 & 4)[b]					
Panel Grade	Common Nail Size[f]				6	6	4	3		
APA STRUCTURAL I grades	6d[g] (0.113" dia.)	1-1/4	5/16	2	185	250	375	420	165	125
				3	210	280	420	475	185	140
	8d (0.131" dia.)	1-3/8	3/8	2	270	360	530	600	240	180
				3	300	400	600	675	265	200

Short Direction

The.Demand

Calculations	NS	Short direction	
longer edge	L	188.0	ft
Total shear of floor	V	116.9	kip
	W	622.0	plf
	R	58465.0	V/2 lbf
shorter edge	H	66.0	ft
flow(DEMAND)	Shear	885.8	plf

Design

Group	Lable	Capcity Vxi
1	X0	1290
2	X1	530
3	X2	360
4	X3	240

	Group	Lable	Capcity Vxi	Plywood t	Min framing b	Continuous edges	Other Edges	Blocked edges	Capacity(plf)	Nail size		APA	
	1	X0	1290	15/32	2(2x)	2 1/2	3	Y	1290	10d	3	2 1/2	2(2x)
	2	X1	530	3/8	2x	2 1/2	4	Y	530	8d	4	2 1/2	2x
	3	X2	360	3/8	2x	6	4	Y	360	8d	4	6	2x
no blck case 1	4	X3	240	3/8	2x	6	0	N	240	8d	0	6	2x

USE

Group	from X=	to X=
1	0	37.8
2	37.8	55.8
3	55.8	68.5
4	68.5	94.0

symmetric for both sides

Long Direction

Long   span

Calculations		NS	Short direction	
shorter edge		L	66.0	ft
Total shear of floor		V	116.9	kip
		W	1771.7	plf
		R	58465.0	V/2 lbf
Longer edge		H	188.0	ft
flow(DEMAND)		Shear	311.0	plf

Group		Capcity Vxi
3	Y0	360
4'	Y1	180

Group	Lable	Capcity Vxi	Plywood t	Min framing b	Continuous edges	Other Edges	Blocked edges	Capacity(plf)	Nail size		APA	
3	Y0	360	3/8	2x	6	4	Y	360	8d	4	6	2x
4'	Y1	180	3/8	2x	6	0	N	180	8d	0	6	2x

USE

Group	from Y=	to Y=
3	0	13.9
4'	13.9	33.0

Nailing layout:

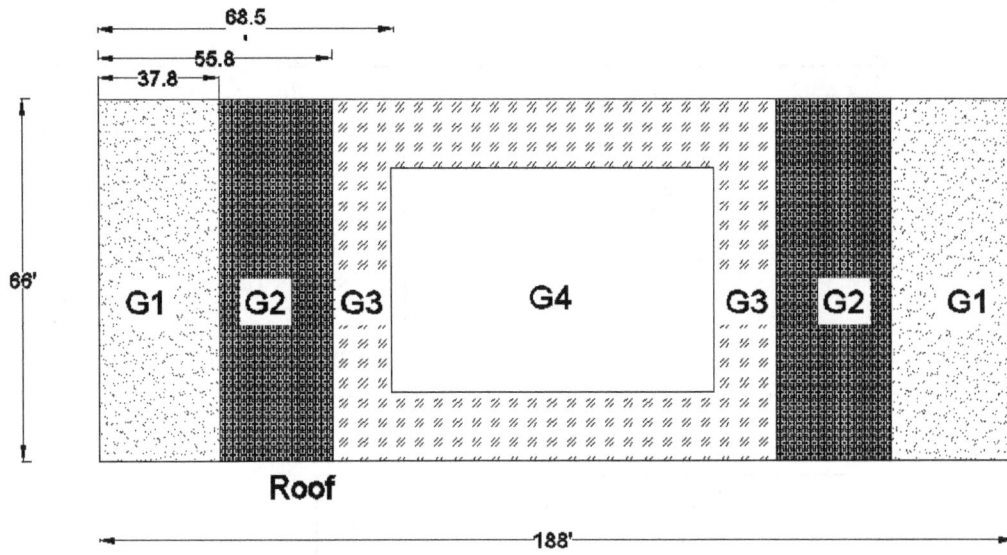

COVER PAGE : FLOOR SHEAR WALL NAILING DESIGN

Design of Shear Wall Nailing :   (Floor)

Summary :

PLAN

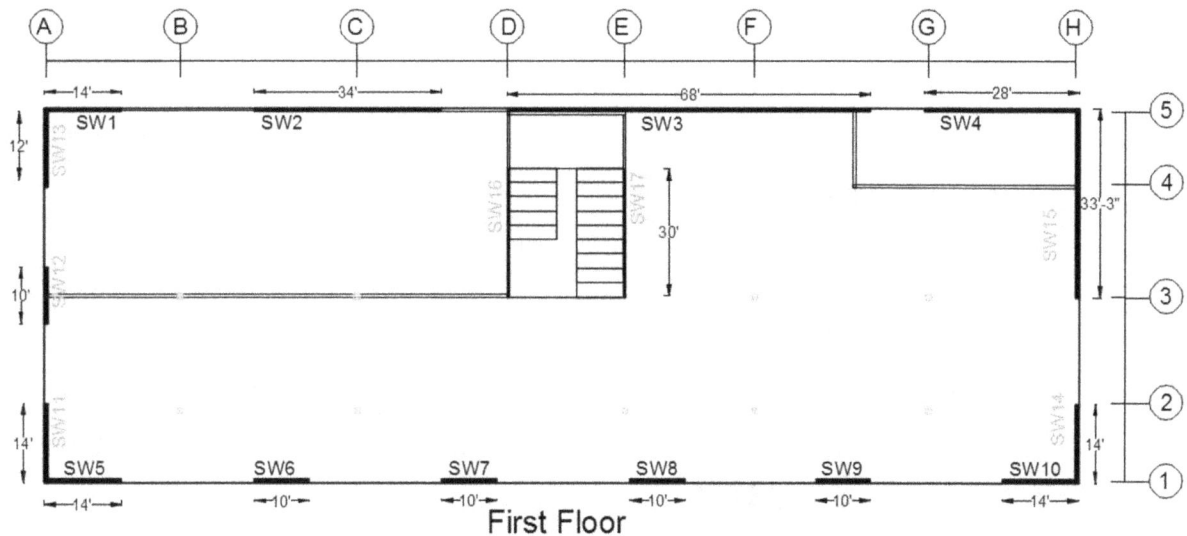

First Floor

Design:

FLOOR						
Shear Wall Line	A	H	D	E	1	5
Shear Base Force (kip)	182.41	182.41	182.41	182.41	182.41	182.41
Shear Wall Reaction Force(kip)	45.6025	45.6025	45.6025	45.6025	91.205	91.205
Wall L (ft)	36	47	30	30	68	144
Reduced shear ratio	1	1	1	1	1	1
Shear Flow (plf)	1266.7	963.5	1520.1	1520.1	1341.3	633.4
Modified shear flow (plf)	1266.7	963.5	1520.1	1520.1	1341.3	633.4
Plywood Type	Double(15/32)	Double(15/32)	Double(15/32)	Double(15/32)	Double(15/32)	15/32
Nail	10d	10d	10d	10d	10d	10d
Nail Space	2	2	2	2	2	2
Capcity (plf)	1740	1740	1740	1740	1740	870
Check	OK	OK	OK	OK	OK	OK

The Purpose of this section is designing the nailing for shear walls belonging to Floor level. The first step is to show the height of Wall; It is the height from the floor to bottom of Glulam Beam. For this building this height is 14 ft. as shown below:

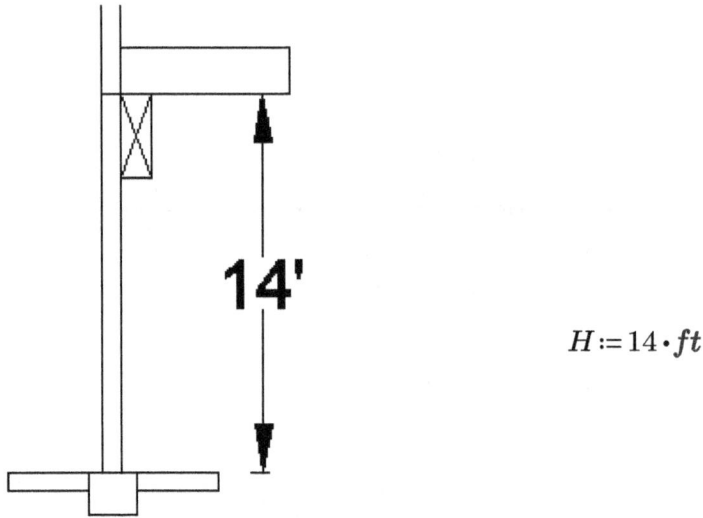

14'

$$H := 14 \cdot ft$$

Wall Section

First Floor Plan:

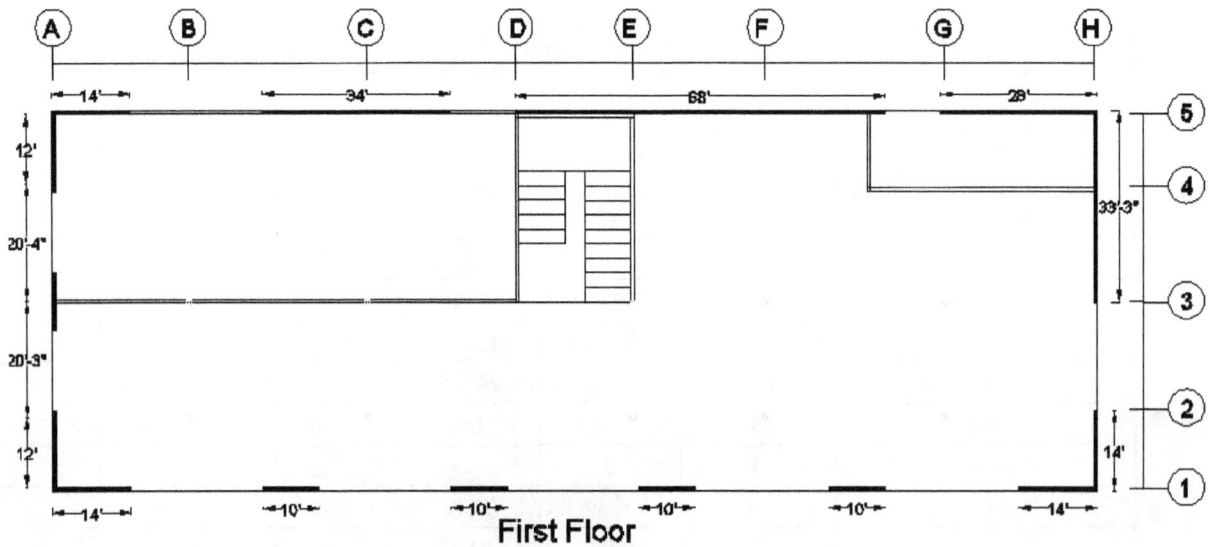

First Floor

Labling the shear Walls:

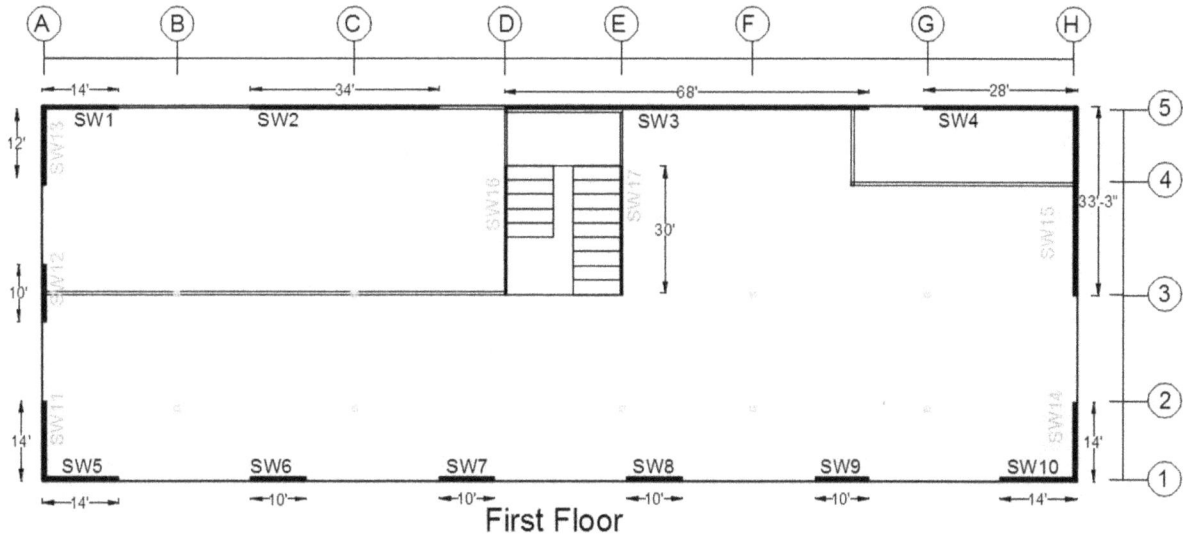

First Floor

The Specifications for each line of shear walls is mentioned below:

FLOOR				Shear Walls along A	
Shear Wall Height, h(ft)	16				
Wall No.	11	12	13	REDUCED VALUE	
Shear Wall length, W (ft)	12	10	14	1	
h/W ratio	1.33	1.60	1.14	TOTAL L(ft)	
Reduced shear value	1.00	1.00	1.00	36	
Total LSW (ft)	12	10	14		

Shear Walls along H

FLOOR		
Shear Wall Height, h(ft)	16	
Wall No.	14	15
Shear Wall length,W (ft)	33.33	14
h/W ratio	0.48	1.14
Reduced shear value	1.00	1.00
Total LSW (ft)	33	14

REDUCED VALUE
1
TOTAL L(ft)
47

Shear Walls along 1

FLOOR							REDUCED VALUE
Shear Wall Height, h(ft)	16						
Wall No.	5	6	7	8	9	10	1
Shear Wall length,W (ft)	14	10	10	10	10	14	TOTAL L(ft)
h/W ratio	1.14	1.60	1.60	1.60	1.60	1.14	68
Reduced shear value	1.00	1.00	1.00	1.00	1.00	1.00	
Total LSW (ft)	14	10	10	10	10	14	

Shear Walls along 5

FLOOR				
Shear Wall Height, h(ft)	16			
Wall No.	1	2	3	4
Shear Wall length,W (ft)	14	34	68	28
h/W ratio	1.14	0.47	0.24	0.57
Reduced shear value	1.00	1.00	1.00	2.00
Total LSW (ft)	14	34	68	28

REDUCED VALUE
1
TOTAL L(ft)
144

FLOOR		Shear Walls along D
Shear Wall Height, h(ft)	16	
Wall No.	12	
Shear Wall length, W (ft)	30	
h/W ratio	0.53	
Reduced shear value	1.00	
Total LSW (ft)	30	

REDUCED VALUE
1
TOTAL L(ft)
30

FLOOR		Shear Walls along E
Shear Wall Height, h(ft)	16	
Wall No.	13	
Shear Wall length, W (ft)	30	
h/W ratio	0.53	
Reduced shear value	1.00	
Total LSW (ft)	30	

REDUCED VALUE
1
TOTAL L(ft)
30

W/H ratio of the shear wall has to be less than one if it is between 2 and 3.5 the reduction factor is applied to that. For this floor, all of the reduction factors are 1.

DESIGN

FLOOR						
Shear Wall Line	A	H	D	E	1	5
Shear Base Force (kip)	182.41	182.41	182.41	182.41	182.41	182.41
Shear Wall Reaction Force(kip)	45.6025	45.6025	45.6025	45.6025	91.205	91.205
Wall L (ft)	36	47	30	30	68	144
Reduced shear ratio	1	1	1	1	1	1
Shear Flow (plf)	1266.7	963.5	1520.1	1520.1	1341.3	633.4
Modified shear flow (plf)	1266.7	963.5	1520.1	1520.1	1341.3	633.4
Plywood Type	Double(15/32)	Double(15/32)	Double(15/32)	Double(15/32)	Double(15/32)	15/32
Nail	10d	10d	10d	10d	10d	10d
Nail Space	2	2	2	2	2	2
Capcity (plf)	1740	1740	1740	1740	1740	870
Check	OK	OK	OK	OK	OK	OK

The above table is valid for design according to given shear load.

COVER PAGE : ROOF SHEAR WALL NAILING DESIGN

Design of Shear Wall Nailing :   (ROOF)

Summary :

PLAN

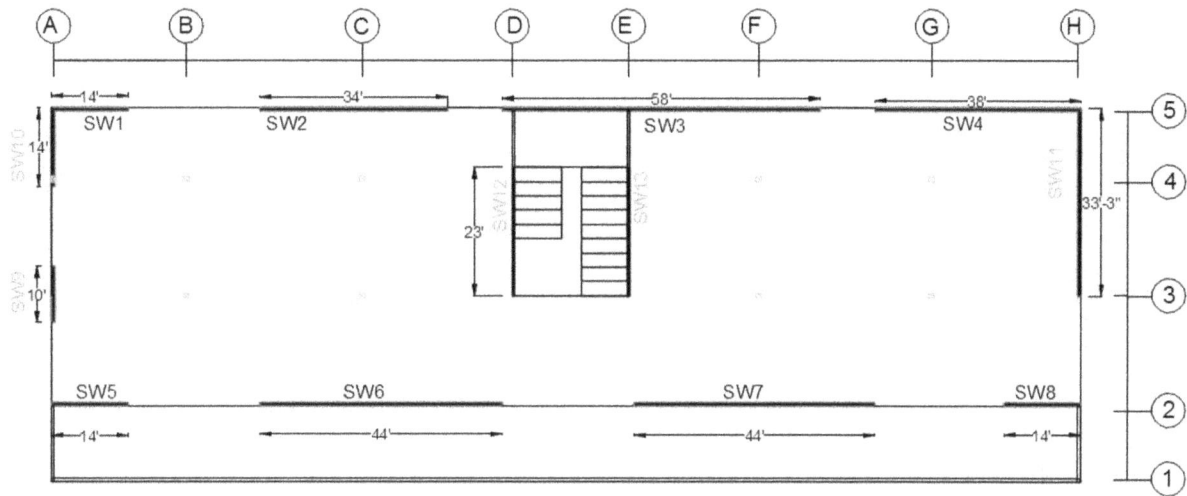

Roof

Design:

FLOOR						
Shear Wall Line	A	H	D	E	2	5
Shear Base Force (kip)	116.93	116.93	116.93	116.93	116.93	116.93
Shear Wall Reaction Force(kip)	29.2325	29.2325	29.2325	29.2325	58.465	58.465
Wall L (ft)	24	33	23	23	116	144
Reduced shear ratio	1	1	1	1	1	1
Shear Flow (plf)	1218.0	877.1	1271.0	1271.0	504.0	406.0
Modified shear flow (plf)	1218.0	877.1	1271.0	1271.0	504.0	406.0
Plywood Type	Double(15/32)	Double(15/32)	Double(15/32)	Double(15/32)	15/32	15/32
Nail	10d	10d	10d	10d	10d	10d
Nail Space	2	3	2	2	4	4
Capcity (plf)	1460	870	1460	1460	510	510
Check	OK	OK	OK	OK	OK	OK

The Purpose of this section is designing the nailing for shear walls belonging to Roof level. The first step is to show the height of Wall; It is the height from the floor to bottom of Glulam Beam. For this building this height is 14 ft. as shown below:

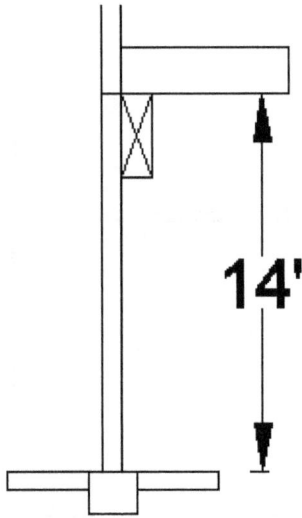

$$H := 14 \cdot ft$$

Wall Section

In next steps, the required nailing and plywood for shear wall are delivered.

Labling the shear Walls:

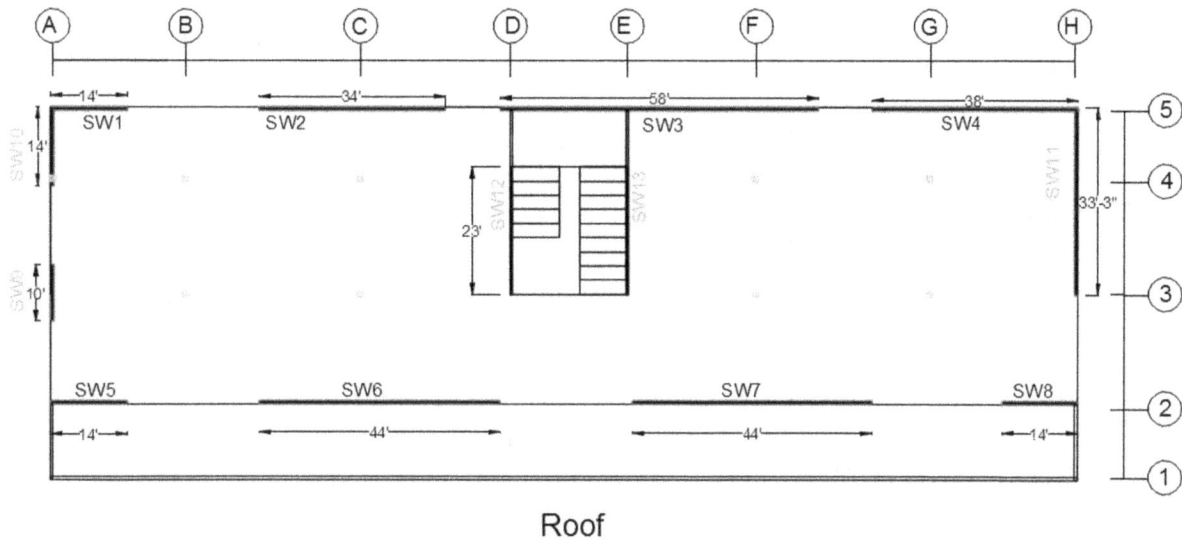

Roof

The Specifications for each line of shear walls is mentioned below:

FLOOR		Shear Walls along A
Shear Wall Height, h(ft)	16	
Wall No.	9	10
Shear Wall length,W (ft)	10	14
h/W ratio	1.60	1.14
Reduced shear value	1.00	1.00
Total LSW (ft)	10	14

REDUCED VALUE
1
TOTAL L(ft)
24

Shear Walls along H

FLOOR	
Shear Wall Height, h(ft)	16
Wall No.	11
Shear Wall length, W (ft)	33.33
h/W ratio	0.48
Reduced shear value	1.00
Total LSW (ft)	33

REDUCED VALUE
1
TOTAL L(ft)
33

Shear Walls along 2

FLOOR				
Shear Wall Height, h(ft)	16			
Wall No.	5	6	7	8
Shear Wall length, W (ft)	14	10	10	10
h/W ratio	1.14	1.60	1.60	1.60
Reduced shear value	1.00	1.00	1.00	1.00
Total LSW (ft)	14	44	44	14

REDUCED VALUE
1
TOTAL L(ft)
116

Shear Walls along 5

FLOOR				
Shear Wall Height, h(ft)	16			
Wall No.	1	2	3	4
Shear Wall length, W (ft)	14	34	58	38
h/W ratio	1.14	0.47	0.28	0.42
Reduced shear value	1.00	1.00	1.00	2.00
Total LSW (ft)	14	34	58	38

REDUCED VALUE
1
TOTAL L(ft)
144

FLOOR		Shear Walls along D
Shear Wall Height, h(ft)	16	
Wall No.	12	
Shear Wall length, W (ft)	23	
h/W ratio	0.70	
Reduced shear value	1.00	
Total LSW (ft)	23	

REDUCED VALUE
1
TOTAL L(ft)
23

FLOOR		Shear Walls along E
Shear Wall Height, h(ft)	16	
Wall No.	13	
Shear Wall length, W (ft)	23	
h/W ratio	0.70	
Reduced shear value	1.00	
Total LSW (ft)	23	

REDUCED VALUE
1
TOTAL L(ft)
23

W/H ratio of the shear wall has to be less than one. If it is between 2 and 3.5 the reduction factor is applied to that. For this floor, all of the reduction factors are 1. Because of high intensity of shear flow, the double plywood is used for most of the cases. Two additional shear walls are located on the sides of the staircase.

DESIGN

FLOOR						
Shear Wall Line	A	H	D	E	2	5
Shear Base Force (kip)	116.93	116.93	116.93	116.93	116.93	116.93
Shear Wall Reaction Force(kip)	29.2325	29.2325	29.2325	29.2325	58.465	58.465
Wall L (ft)	24	33	23	23	116	144
Reduced shear ratio	1	1	1	1	1	1
Shear Flow (plf)	1218.0	877.1	1271.0	1271.0	504.0	406.0
Modified shear flow (plf)	1218.0	877.1	1271.0	1271.0	504.0	406.0
Plywood Type	Double(15/32)	Double(15/32)	Double(15/32)	Double(15/32)	15/32	15/32
Nail	10d	10d	10d	10d	10d	10d
Nail Space	2	3	2	2	4	4
Capcity (plf)	1460	870	1460	1460	510	510
Check	OK	OK	OK	OK	OK	OK

The above table is valid for design according to given shear load.

www.ingramcontent.com/pod-product-compliance
Lightning Source LLC
Chambersburg PA
CBHW080526220326
41599CB00032B/6212